Study Guide/ Student Solutions Manual to accompany

Principles of Physics
THIRD EDITION
A CALCULUS-BASED TEXT VOLUME 1
Serway & Jewett

John R. Gordon
James Madison University

Ralph McGrew
Broome Community College

Raymond A. Serway
James Madison University

John W. Jewett, Jr.
California State Polytechnic University-Pomona

BROOKS/COLE
THOMSON LEARNING

Australia • Canada • Mexico • Singapore • Spain
United Kingdom • United States

**For more information about our products,
contact us at:
Thomson Learning Academic Resource Center
1-800-423-0563**

**For permission to use material from this text,
contact us by:
Phone: 1-800-730-2214
Fax: 1-800-731-2215
Web: www.thomsonrights.com**

Asia
Thomson Learning
60 Albert Complex, #15-01
Alpert Complex
Singapore 189969

Australia
Nelson Thomson Learning
102 Dodds Street
South Street
South Melbourne, Victoria 3205
Australia

Canada
Nelson Thomson Learning
1120 Birchmount Road
Toronto, Ontario M1K 5G4
Canada

Europe/Middle East/South Africa
Thomson Learning
Berkshire House
168-173 High Holborn
London WC1 V7AA
United Kingdom

Latin America
Thomson Learning
Seneca, 53
Colonia Polanco
11560 Mexico D.F.
Mexico

Spain
Paraninfo Thomson Learning
Calle/Magallanes, 25
28015 Madrid, Spain

Table of Contents

Preface

This <u>Student Solutions Manual and Study Guide</u> has been written to accompany the textbook **Principles of Physics**, Third Edition, Volume I, by Raymond A. Serway and John Jewett. The purpose of this <u>Student Solutions Manual and Study Guide</u> is to provide students with a convenient review of the basic concepts and applications presented in the textbook, together with solutions to selected end-of-chapter problems. This is not an attempt to rewrite the textbook in a condensed fashion. Rather, emphasis is placed upon clarifying typical troublesome points, and providing further practice in methods of problem solving.

Each chapter is divided into several parts, and every textbook chapter has a matching chapter in this book. Very often, reference is made to specific equations or figures in the textbook. Each feature of this Study Guide has been included to insure that it serves as a useful supplement to the textbook. Most chapters contain the following components:

- **Notes From Selected Chapter Sections:** This is a summary of important concepts, newly defined physical quantities, and rules governing their behavior.

- **Equations and Concepts:** This is a review of the chapter, with emphasis on highlighting important concepts and describing important equations and formalisms.

- **Suggestions, Skills, and Strategies:** This offers hints and strategies for solving typical problems that the student will often encounter in the course. In some sections, suggestions are made concerning mathematical skills that are necessary in the analysis of problems.

- **Review Checklist:** This is a list of topics and techniques the student should master after reading the chapter and working the assigned problems.

- **Answers to Selected Conceptual Questions:** Suggested responses are provided for twenty percent of the Conceptual Questions.

- **Solutions to Selected End-of-Chapter Problems:** Solutions are given for approximately half of the odd-numbered problems from the text. Problems were selected to illustrate important concepts in each chapter.

- **Tables:** A list of selected Physical Constants is printed on the inside front cover; and a table of some Conversion Factors is provided on the inside back cover.

A note concerning significant figures: In nearly all problem statements, data is given to three significant figures. The answers to end-of-chapter problems are stated to three significant figures, though intermediate calculation steps are carried out with as many digits as possible. The last digit is uncertain, often depending on the precision of the values assumed for physical constants and properties.

We sincerely hope that this <u>Student Solutions Manual and Study Guide</u> will be useful to you in reviewing the material presented in the text, and in improving your ability to solve problems and score well on exams. We welcome any comments or suggestions which could help improve the content of this study guide in future editions; and we wish you success in your study.

John R. Gordon
Dept. of Physics
James Madison University
Harrisonburg, VA 22807

Ralph McGrew
Dept. of Engineering Science, AT-101
Broome Community College
Binghamton, NY 13902-1017

Raymond A. Serway

John W. Jewett, Jr.

Acknowledgments

It is a pleasure to acknowledge the excellent work of Michael Rudmin of DSC Publishing whose attention to detail in the preparation of the camera-ready copy did much to enhance the quality of this edition of the <u>Student Solutions Manual and Study Guide</u>. His graphics skills and technical expertise combined to produce illustrations for the first edition which continue to add much to the appearance and usefulness of this volume.

Special thanks for managing all phases of this project go to the editorial staff at Saunders College Publishing: our Developmental Editor Ed Dodd, our Technology Editor Ken McFarlane, and our Senior Production Manager Charlene Squib. Finally, we express our appreciation to our families for their inspiration, patience, and encouragement.

Suggestions for Study

Very often we are asked "How should I study this subject, and prepare for examinations?" There is no simple answer to this question, however, we would like to offer some suggestions which may be useful to you.

1. It is essential that you understand the basic concepts and principles before attempting to solve assigned problems. This is best accomplished through a careful reading of the textbook before attending your lecture on that material, jotting down certain points which are not clear to you, taking careful notes in class, and asking questions. You should reduce memorization of material to a minimum. Memorizing sections of a text, equations, and derivations does not necessarily mean you understand the material. Perhaps the best test of your understanding of the material will be your ability to solve the problems in the text, or those given on exams.

2. Try to solve as many problems at the end of the chapter as possible. You will be able to check the accuracy of your calculations to the odd-numbered problems, since the answers to these are given at the back of the text. Furthermore, detailed solutions to approximately half of the odd-numbered problems are provided in this study guide. Many of the worked examples in the text will serve as a basis for your study.

3. The method of solving problems should be carefully planned. First, read the problem several times until you are confident you understand what is being asked. Look for key words which will help simplify the problem, and perhaps allow you to make certain assumptions. You should also pay special attention to the information provided in the problem. In many cases a simple diagram is a good starting point; and it is always a good idea to write down the given information before trying to solve the problem. After you have decided on the method you feel is appropriate for the problem, proceed with your solution. If you are having difficulty in working problems, we suggest that you again read the text and your lecture notes. It may take several readings before you are ready to solve certain problems, though the solved problems in this Study Guide should be of value to you in this regard. However, your solution to a problem does not have to look just like the one presented here. A problem can sometimes be solved in different ways, starting from different principles. If you wonder about the validity of an alternative approach, ask your instructor.

4. After reading a chapter, you should be able to define any new quantities that were introduced, and discuss the first principles that were used to derive fundamental formulas. A review is provided in each chapter of the Study Guide for this purpose, and the marginal notes in the textbook (or the index) will help you locate these topics. You should be able to correctly associate with each physical quantity the symbol used to represent that quantity (including vector notation, if appropriate) and the SI unit in which the quantity is specified. Furthermore, you should be able to express each important formula or equation in a concise and accurate prose statement.

5. We suggest that you use this Study Guide to review the material covered in the text, and as a guide in preparing for exams. You should also use the Chapter Review, Notes From Selected Chapter Sections, and Equations and Concepts to focus in on any points which require further study. Remember that the main purpose of this Study Guide is to improve upon the efficiency and effectiveness of your study hours and your overall understanding of physical concepts. However, it should not be regarded as a substitute for your textbook or individual study and practice in problem solving.

Chapter 1

Introduction and Vectors

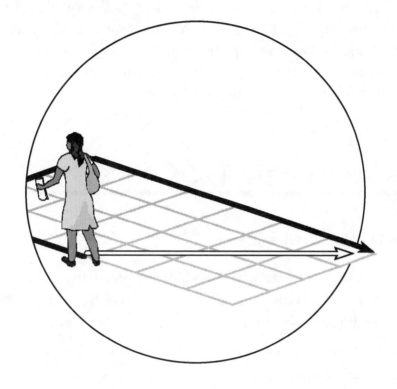

INTRODUCTION

The goal of physics is to provide an understanding of nature by developing theories based on experiments. The theories are usually expressed in mathematical form. Fortunately, it is possible to explain the behavior of a variety of physical systems with a limited number of fundamental laws.

Since the following chapters will be concerned with the laws of physics, we must begin by clearly defining the basic quantities involved in these laws. For example, such physical quantities as force, velocity, volume, and acceleration can be described in terms of more fundamental quantities. In the next several chapters we shall encounter three basic quantities: length (L), mass (M), and time (T). In later chapters we will need to add two other basic units to our list, for temperature (the Kelvin) and for electric current (the ampere). In our study of mechanics, however, we shall be concerned only with the units of length, mass and time.

Physical quantities that have both numerical and directional properties are represented by vectors. Some examples of vector quantities are force, displacement, velocity, and acceleration. This chapter includes a discussion of vector algebra and some general properties of vector quantities. The addition and subtraction of vector quantities are discussed, together with some common applications to physical situations.

Vector quantities are used throughout this text, and it is therefore imperative that you master both their graphical and their algebraic properties.

NOTES FROM SELECTED CHAPTER SECTIONS

1.1   Standards of Length, Mass, and Time

Mechanical quantities can be expressed in terms of three fundamental quantities, **mass, length,** and **time,** which in the SI system have the units **kilograms** (kg), **meters** (m), and **seconds** (s), respectively.

1.2   Density and Atomic Mass

The density of a substance is defined as its **mass per unit volume**. Different substances have different densities mainly because of differences in their atomic masses and atomic arrangements.

The masses of atomic and nuclear particles are expressed in terms of **atomic mass units** (u). The mass of one ^{12}C atom is defined to be exactly 12 u.

$$1 \text{ u} = 1.661 \times 10^{-27} \text{ kg}$$
$$\text{mass of proton} = 1.0073 \text{ u}$$
$$\text{mass of neutron} = 1.0087 \text{ u}$$

1.3   Dimensional Analysis

It is often useful to use the **method of dimensional analysis** to check equations and to assist in deriving expressions. Dimensional analysis makes use of the fact that **dimensions can be treated as algebraic quantities**. That is, quantities can be added or subtracted only if they have the same dimensions. Furthermore, the terms on both sides of an equation must have the same dimensions. By following these simple rules, you can use dimensional analysis to help determine whether or not an expression has the correct form, because the relationship can be correct only if the dimensions on the two sides of the equation are the same.

1.6 Significant Figures

When one performs measurements on certain quantities, the accuracy of the measured values can vary; that is, the true values are known only to be within the limits of the experimental uncertainty. The value of the uncertainty can depend on various factors such as the quality of the apparatus, the skill of the experimenter, and the number of measurements performed.

When multiplying several quantities, the number of significant figures in the final answer is the same as the number of significant figures in the **least** accurate of the quantities being multiplied, where "least accurate" means "having the lowest number of significant figures." The same rule applies to division.

When numbers are added (or subtracted), the number of decimal places in the result should equal the smallest number of decimal places of any term in the sum.

1.7 Coordinate Systems

The location of any point in space can be specified by the use of a coordinate system. Two frequently used coordinate systems are the **Cartesian** (or **rectangular**) coordinate system and the **plane polar coordinate** system.

In general, any coordinate system consists of:

- A fixed reference point O, called the origin,
- A set of specified axes or directions with an appropriate scale and axis labels.
- Instructions that tell us how to label a point in space relative to the origin and axes.

1.8 Vectors and Scalars

A **vector** is a physical quantity that must be specified by both magnitude and direction.

A **scalar** quantity has only magnitude.

1.9   Some Properties of Vectors

- Equality of Two Vectors—Two vectors are **equal** if they have the same magnitude and the same direction.
- Addition of Vectors—In order for two or more vectors to be added, they must have the same units; their sum is independent of the order of addition. The triangle method and the parallelogram method are graphic methods for determining the resultant or sum of two or more vectors.
- Negative of a Vector—The sum of a vector and its negative is zero. A vector and its negative have the same magnitude, but have opposite directions.
- Multiplication by a Scalar—When a vector is **multiplied** or **divided** by a positive (negative) scalar, the result is a vector in the same (opposite) direction. The magnitude of the resulting vector is equal to the product of the absolute value of the scalar and the magnitude of the original vector.

1.10   Components of a Vector and Unit Vectors

Any vector can be completely described by its components. A **unit vector** is a dimensionless vector **with a magnitude of one** used to **specify a given direction**.

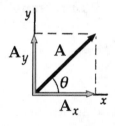

Figure 1.1 Components of a vector

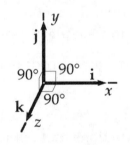

Figure 1.2 Unit Vectors

The unit vectors **i**, **j**, and **k** form a set of **mutually perpendicular** vectors as illustrated above.

$$\mathbf{i} \equiv \text{a unit vector along the } x \text{ axis}$$
$$\mathbf{j} \equiv \text{a unit vector along the } y \text{ axis}$$
$$\mathbf{k} \equiv \text{a unit vector along the } z \text{ axis}$$
$$\text{where } |\mathbf{i}| = |\mathbf{j}| = |\mathbf{k}| = 1$$

Mathematical Notation

It is often convenient to use symbols to represent mathematical operations and/or relationships between or among variables or values of physical quantities. Some symbols which will be used throughout your textbook and in the Solutions Manual are shown below.

Symbol	Meaning
$=$	equality of two quantities
$\approx$	approximately equal to
$\sim$	on the order of
$\equiv$	defined as
$\propto$	proportional to
$<$	less than
$>$	greater than
Δ	change in a quantity
$\lvert x \rvert$	magnitude of the quantity x
$\sum x_i$	sum of the set x

EQUATIONS AND CONCEPTS

The density of any substance is defined as the ratio of mass to volume.

$$\rho \equiv \frac{m}{V} \tag{1.1}$$

The location of a point P in a plane can be specified by either Cartesian coordinates, x and y, or polar coordinates, r and θ. If one set of coordinates is known, values for the other set can be calculated.

$$x = r\cos\theta \tag{1.2}$$

$$y = r\sin\theta \tag{1.3}$$

$$\tan\theta = \frac{y}{x} \tag{1.4}$$

$$r = \sqrt{x^2 + y^2} \tag{1.5}$$

Vector quantities obey the **commutative law of addition.** In order to add vector **A** to vector **B** using the graphical method, first construct **A**, and then draw **B** such that the tail of **B** starts at the head of **A**. The sum of **A** + **B** is the vector that completes the triangle by connecting the tail of **A** to the head of **B**. See Figure 1.3(a).

$$\mathbf{A} + \mathbf{B} = \mathbf{B} + \mathbf{A} \tag{1.7}$$

When three or more vectors are added, the sum is independent of the manner in which the vectors are grouped. This is the **associative law** of addition.

$$\mathbf{A} + (\mathbf{B} + \mathbf{C}) = (\mathbf{A} + \mathbf{B}) + \mathbf{C} \tag{1.8}$$

When more than two vectors are to be added, they are all connected head-to-tail in any order. The resultant or sum is the vector which joins the tail of the first vector to the head of the last vector. See Figure 1.3(b).

$$\mathbf{R} = \mathbf{A} + \mathbf{B} + \mathbf{C} + \mathbf{D}$$

When two or more vectors are to be added, all of them must represent the same physical quantity—that is, have the same units. In the graphical or geometrical method of vector addition, the length of each vector is proportional to the magnitude of the vector. Also, each vector must be pointed along a direction which makes the proper angle relative to the others.

The operation of vector subtraction utilizes the definition of the negative of a vector. The negative of vector **A** is the vector which has a magnitude equal to the magnitude of **A**, but acts or points along a direction opposite the direction of **A**. See Figure 1.3(c)

$$\mathbf{A} - \mathbf{B} = \mathbf{A} + (-\mathbf{B}) \qquad (1.9)$$

A vector **A** in a two-dimensional coordinate system can be resolved into its components along the x and y directions. The projection of **A** onto the x axis is the x component of **A**; and the projection of **A** onto the y axis is the y component of **A**.

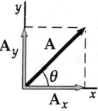

The magnitude of **A** and the angle, θ, which the vector makes with the positive x axis can be determined from the values of the x and y components of **A**.

$$A_x = A\cos\theta \qquad (1.10)$$

$$A_y = A\sin\theta$$

$$A = \sqrt{A_x^2 + A_y^2} \qquad (1.11)$$

$$\tan\theta = \frac{A_y}{A_x} \qquad (1.12)$$

A vector **A** lying in the x-y plane, having rectangular components A_x and A_y, can be expressed in unit vector notation.

$$\mathbf{A} = A_x\mathbf{i} + A_y\mathbf{j} \qquad (1.13)$$

When two vectors are added, the resultant vector can be expressed in terms of the components of the two vectors.

If $\quad \mathbf{R} = \mathbf{A} + \mathbf{B}: \qquad (1.14)$

$$\mathbf{R} = (A_x + B_x)\mathbf{i} + (A_y + B_y)\mathbf{j}$$

The components of the resultant vector can be expressed in terms of the components of the two vectors to be added.

$$R_x = A_x + B_x$$
$$R_y = A_y + B_y \qquad (1.15)$$

The magnitude of the resultant vector and the direction it makes with the positive x-axis can be obtained from its components.

$$R = \sqrt{R_x{}^2 + R_y{}^2} \qquad (1.16)$$

$$\tan\theta = \frac{R_y}{R_x} = \frac{A_y + B_y}{A_x + B_x} \qquad (1.17)$$

SUGGESTIONS, SKILLS, AND STRATEGIES

Give careful study to Section 1.11, Modeling, Alternative Representations, and Problem-Solving Strategy, on pages 25-30 of your textbook. Forming appropriate **models** and forming **alternative representations** are important steps in successful problem solving. You will use **geometric models** immediately in chapter 1 and throughout the rest of your course. Return to Section 1.11 to review the framework it presents as you progress through your course. During the whole course you will learn to use particular **simplification models, analysis**

models, and **structural models** in solving physics problems. You will choose and translate among **mental, pictorial, simplified pictorial, graphical, tabular, and mathematical representations** as ways of viewing or presenting the information related to the problem. Page 29 of the textbook presents a five-step general problem-solving strategy which you will likely find essential.

You should be familiar with some important mathematical techniques:

- Using powers of ten in expressing such numbers as $0.00058 = 5.8 \times 10^{-4}$.

- Basic algebraic operations such as factoring, handling fractions, solving quadratic equations, and solving linear equations.

- The fundamentals of plane and solid geometry—including the ability to graph functions, calculate the areas and volumes of standard geometric figures and recognize the equations and graphs of a straight line, a circle, an ellipse, a parabola and a hyperbola.

- The basic ideas of trigonometry—definitions and properties of the sine, cosine, and tangent functions; the Pythagorean Theorem, the law of cosines, the law of sines, and some of the basic trigonometric identities.

Adding vectors

When two or more vectors are to be added, the following step-by-step procedure is recommended:

- Select a coordinate system.

- Draw a sketch of the vectors to be added (or subtracted), with a label on each vector.

- Find the x and y components of all vectors.

- Find the resultant components (the algebraic sum of the components) in both the x and y directions.

- Use the Pythagorean theorem to find the magnitude of the resultant vector.

- Use a suitable trigonometric function to find the angle the resultant vector makes with the x axis.

To add vector **A** to vector **B** graphically, first construct **A**, and then draw **B** such that the tail of **B** starts at the head of **A**. The sum **A + B** is the vector that completes the triangle as shown in Figure 1.3(a) below. The procedure for adding more than two vectors (the polygon rule) is also illustrated.

In order to subtract two vectors **graphically**, recognize that **A − B** is equivalent to the operation **A + (−B)**. Since the vector **−B** is a vector whose magnitude is B and is opposite in direction to **B**, the construction in Figure 1.3 (c) below is obtained:

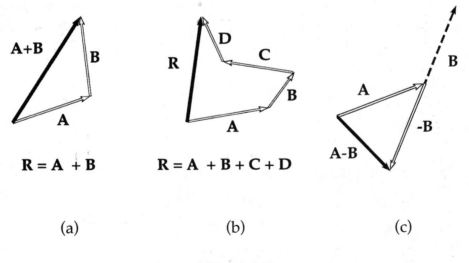

$R = A + B$	$R = A + B + C + D$	
(a)	(b)	(c)

Figure 1.3

Adding vectors by (a) the triangle rule and (b) the polygon rule.
(c) Subtracting two vectors graphically.

REVIEW CHECKLIST

▷   Discuss the units of length, mass and time and the standards for these quantities in SI units; and perform a **dimensional analysis** of an equation containing physical quantities whose individual units are known.

▷   **Convert units** from one system to another.

▷   Become familiar with the meaning of various mathematical symbols and Greek letters. Identify and properly use mathematical notations such as the following: ∝ (is proportional to), < (is less than), ≈ (is approximately equal to), and Δ (a change in quantity).

▷   Locate a point in space using both Cartesian coordinates and polar coordinates.

▷   Describe the basic properties of vectors such as the rules of vector addition and graphical solutions for the addition of two or more vectors.

▷   Resolve a vector into its rectangular components. Determine the magnitude and direction of a vector from its rectangular components.

▷   Understand the use of unit vectors to express any vector in unit vector notation.

ANSWERS TO SELECTED CONCEPTUAL QUESTIONS

1.   Suppose that the three fundamental standards of the metric system were length, **density**, and time rather than length, **mass**, and time. The standard of density in this system is to be defined as that of water. What considerations about water would need to be addressed to make sure that the standard unit of density is as accurate as possible?

Answer   A number of considerations would be necessary. There are the environmental details related to the water—a standard temperature would have to be defined, as well as a standard pressure. Another consideration is the quality of the water, in terms of defining a lower limit of impurities. Another problem with this scheme is that density cannot be measured directly with a single measurement, as can length, mass and time. As a combination of two measurements (mass, and volume, which itself involves **three** measurements!), it has higher inherent uncertainty than a single measurement.

□   □   □   □

4. Express the following quantities using the prefixes given in Table 1.4: (a) 3×10^{-4} m, (b) 5×10^{-5} s, (c) 72×10^{2} g.

Answer　　(a) 0.300 mm (b) 50.0 μs (c) 7.20 kg.

Power	Prefix	Abbreviation
−24	yocto	y
−21	zepto	z
−18	atto	a
−15	femto	f
−12	pico	p
−9	nano	n
−6	micro	μ
−3	milli	m
−2	centi	c
−1	deci	d
3	kilo	k
6	mega	M
9	giga	G
12	tera	T
15	peta	P
18	exa	E
21	zetta	Z
24	yotta	Y

Table 1.4 Prefixes for powers of ten

□　□　□　□

11. A vector **A** lies in the *xy* plane. For what orientations of **A** will both of its components be negative? For what orientations will its components have opposite signs?

Answer　　The vector **A** will have both rectangular components negative in quadrant III, when the angle of **A** is between π rad (180°) and $3\pi/2$ rad (270°). The vector **A** will have components with opposite signs in two cases: first, in quadrant II, when the angle of **A** is between $\pi/2$ rad (90°) and π rad (180°), and second, in quadrant IV, when the angle of **A** is between $3\pi/2$ rad (270°) and 2π rad (360°).

□　□　□　□

16. Is it possible to add a vector quantity to a scalar quantity? Explain.

Answer　　Vectors and scalars are distinctly different and cannot be added to each other. Remember that a vector defines a quantity **in a certain direction**, while a scalar only defines a quantity with no associated direction.

SOLUTIONS TO SELECTED END-OF-CHAPTER PROBLEMS

9.   The position of a particle when moving under uniform acceleration is some function of the elapsed time and the acceleration. Suppose we write this position as $s = k a^m t^n$, where k is a dimensionless constant. Show by dimensional analysis that this expression is satisfied if $m = 1$ and $n = 2$. Can this analysis give the value of k?

Solution

For the equation to be valid, we must choose values of m and n to make it dimensionally consistent. Since s is a position, its dimensions are those of length (L). The acceleration, a, is a length divided by the square of a time (L/T^2). The variable t has dimensions of time (T), and the constant k has no dimensions. Substituting these dimensions into the equation yields:

$$(L) = \left(\frac{L}{T^2}\right)^m (T)^n = (L)^m (T)^{-2m} (T)^n \qquad \text{or} \qquad (L)^1 (T)^0 = (L)^m (T)^{n-2m}$$

Note that the factor $(T)^0$ introduced on the left side of the second equation is equal to 1. This equation can be true only if the powers of length (L) are the same on the two sides of the equation and, simultaneously, the powers of time (T) are the same on both sides. Indeed, if another basic unit such as mass (M) were present, we would also require that its powers be identical on the two sides. Thus, we obtain a set of two simultaneous equations: $1 = m$ and $0 = n - 2m$. The solutions are therefore seen to be: $m = 1$ and $n = 2$ ◊

This gives no information about possible values of the dimensionless constant k.   ◊

---

17.   One gallon of paint (volume = 3.78×10^{-3} m³) covers an area of 25.0 m². What is the thickness of the paint on the wall?

Solution   We assume the paint keeps the same volume in the can and on the wall. We model the film on the wall as a rectangular solid, with its volume given by its surface area multiplied by its uniform thickness t: $V = At$

Therefore, $t = \dfrac{V}{A} = \dfrac{3.78 \times 10^{-3} \text{ m}^3}{25.0 \text{ m}^2} = 1.51 \times 10^{-4}$ m   ◊

19. One cubic meter (1.00 m^3) of aluminum has a mass of 2.70×10^3 kg, and 1.00 m^3 of iron has a mass of 7.86×10^3 kg. Find the radius of a solid aluminum sphere that will balance a solid iron sphere of radius 2.00 cm on an equal-arm balance.

Solution   We require equal masses: $m_{Al} = m_{Fe}$ or $\rho_{Al} V_{Al} = \rho_{Fe} V_{Fe}$

Therefore, $\rho_{Al}\left(\frac{4}{3}\pi r^3\right) = \rho_{Fe}\left(\frac{4}{3}\pi(2.00 \text{ cm})^3\right)$

$$r^3 = \left(\frac{\rho_{Fe}}{\rho_{Al}}\right)(2.00 \text{ cm})^3 = \left(\frac{7.86 \text{ kg / m}^3}{2.70 \text{ kg / m}^3}\right)(2.00 \text{ cm})^3 = 23.3 \text{ cm}^3$$

and $r = 2.86 \text{ cm}$ ◊

---

21. Find the order of magnitude of the number of Ping-Pong balls that would fit into a typical-size room (without being crushed). In your solution state the quantities you measure or estimate and the values you take for them.

Solution   Since the volume of a typical room is much larger than a Ping-Pong ball, we should expect that a very large number of balls (maybe a million) could fit in a room. Since we are only asked to find an estimate, we do not need to be too concerned about how the balls are arranged. Therefore, to find the number of balls we can simply divide the volume of an average-size living room (perhaps 15 ft × 20 ft × 8 ft) by the volume of an individual Ping-Pong ball. Using the approximate conversion 1 ft = 30 cm, we find

$$V_{Room} = (15 \text{ ft})(20 \text{ ft})(8 \text{ ft})(30 \text{ cm / ft})^3 \approx 7 \times 10^7 \text{ cm}^3$$

A Ping-Pong ball has a diameter of about 3 cm, so we can estimate its volume as a cube:

$$V_{ball} = (3 \text{ cm})(3 \text{ cm})(3 \text{ cm}) \approx 30 \text{ cm}^3$$

The number of Ping-Pong balls that can fill the room is

$$N \approx \frac{V_{Room}}{V_{ball}} \approx 2 \times 10^6 \text{ balls} \sim 10^6 \text{ balls}$$ ◊

So a typical room can hold about a million Ping-Pong balls. This problem gives us a sense of how large a quantity "a million" really is.

---

31. A fly lands on one wall of a room. The lower left-hand corner of the wall is selected as the origin of a two-dimensional cartesian coordinate system. If the fly is located at the point having coordinates (2.00, 1.00) m, (a) how far is it from the corner of the room? (b) What is its location in polar coordinates?

Solution

(a) Assume the wall is in the *x-y* plane so that the coordinates are $x = 2.00$ m and $y = 1.00$ m; and the fly is located at point *P*. The distance between two points in the x-y plane is

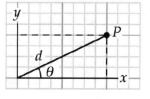

$$d = \sqrt{(x_2 - x_1)^2 + (y_2 - y_1)^2} = \sqrt{(2.00 \text{ m} - 0)^2 + (1.00 \text{ m} - 0)^2} = 2.24 \text{ m} \qquad \Diamond$$

(b) From the figure, $\theta = \tan^{-1}\left(\dfrac{x}{y}\right) = \tan^{-1}\left(\dfrac{1.00}{2.00}\right) = 26.6°$ and $r = d$

Therefore, the polar coordinates of the point *P* are (2.24 m, 26.6°) $\Diamond$

---

35. A skater glides along a circular path of radius 5.00 m. If he coasts around one half of the circle, (a) find the magnitude of the displacement vector and (b) how far the person skated. (c) What is the magnitude of the displacement if he skates all the way around the circle?

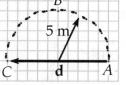

Solution   See the sketch above and to the right.

(a) $|\mathbf{d}| = |-10.0\mathbf{i}| = 10.0$ m $\Diamond$

since the displacement is a straight line from point *A* to point *C*.

(b) The actual distance skated is not equal to the straight-line displacement. The distance follows the curved path of the semicircle (*ABC*).

$$s = \left(\tfrac{1}{2}\right)(2\pi r) = 5.00\pi = 15.7 \text{ m} \qquad \Diamond$$

(c) If the circle is complete, **d** begins and ends at point *A*. Hence, $|\mathbf{d}| = 0$. $\Diamond$

---

37. A roller coaster car moves 200 ft horizontally and then rises 135 ft at an angle of 30.0° above the horizontal. It then travels 135 ft at an angle of 40.0° downward. What is its displacement from its starting point? Use graphical techniques.

Solution Your sketch when drawn to scale should look somewhat like the one to the right. (You will probably only be able to obtain a measurement to one or two significant figures).

The distance R and the angle θ can be measured to give, upon use of your scale factor, the values of:

$R = 4.2 \times 10^2$ ft at about 3° below the horizontal. ◊

———

41. A person going for a walk follows the path shown in Figure P1.41. The total trip consists of four straight-line paths. At the end of the walk, what is the person's resultant displacement measured from the starting point?

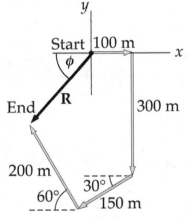

Figure P1.41
(with resultant added)

Solution The resultant displacement $\mathbf{R}$ is equal to the sum of the four individual displacements, $\mathbf{R} = \mathbf{d}_1 + \mathbf{d}_2 + \mathbf{d}_3 + \mathbf{d}_4$

We translate from the pictorial representation to a mathematical representation by writing the individual displacements in unit-vector notation:

$\mathbf{d}_1 = 100\mathbf{i}$ m

$\mathbf{d}_2 = -300\mathbf{j}$ m

$\mathbf{d}_3 = (-150 \cos30°)\mathbf{i}$ m $+ (-150 \sin30°)\mathbf{j}$ m $= -130\mathbf{i}$ m $- 75\mathbf{j}$ m

$\mathbf{d}_4 = (-200 \cos60°)\mathbf{i}$ m $+ (200 \sin60°)\mathbf{j}$ m $= -100\mathbf{i}$ m $+ 173\mathbf{j}$ m

Summing the components together,

$$R_x = d_{1x} + d_{2x} + d_{3x} + d_{4x} = (100 + 0 - 130 - 100) \text{ m} = -130 \text{ m}$$

$$R_y = d_{1y} + d_{2y} + d_{3y} + d_{4y} = (0 - 300 - 75 + 173) \text{ m} = -202 \text{ m}$$

$$\mathbf{R} = -130\mathbf{i} \text{ m} - 202\mathbf{j} \text{ m}$$

◊

$$|\mathbf{R}| = \sqrt{R_x^2 + R_y^2} = \sqrt{(-130 \text{ m})^2 + (-202 \text{ m})^2} = 240 \text{ m}$$

◊

$$\phi = \tan^{-1}\left(\frac{R_y}{R_x}\right) = \tan^{-1}\left(\frac{-202}{-130}\right) = 57.2°$$

Pitfall Prevention 1.9 in the text explains why this angle does not specify the resultant's direction in standard form. Instead, the angle counterclockwise from the +x axis is

$$\theta = 57.2° + 180° = 237°$$

◊

43. Three displacement vectors of a croquet ball are shown in Figure P1.43, where $|\mathbf{A}| = 20.0$ units, $|\mathbf{B}| = 40.0$ units, and $|\mathbf{C}| = 30.0$ units. Find (a) the resultant in unit-vector notation and (b) the magnitude and direction of the resultant vector.

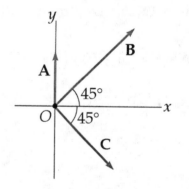

Solution

(a) $A_x = (20.0 \text{ units}) \cos 90° = 0$

$A_y = (20.0 \text{ units}) \sin 90° = 20.0 \text{ units}$

Figure P1.43

$B_x = (40.0 \text{ units}) \cos 45° = 28.3 \text{ units}$

$B_y = (40.0 \text{ units}) \sin 45° = 28.3 \text{ units}$

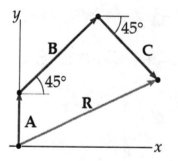

$C_x = (30.0 \text{ units}) \cos 315° = 21.2 \text{ units}$

$C_y = (30.0 \text{ units}) \sin 315° = -21.2 \text{ units}$

$R_x = A_x + B_x + C_x = (0 + 28.3 + 21.2)$ units $= 49.5$ units

$R_y = A_y + B_y + C_y = (20 + 28.3 - 21.2)$ units $= 27.1$ units

$\mathbf{R} = 49.5\mathbf{i} + 27.1\mathbf{j}$

(b) $|\mathbf{R}| = \sqrt{R_x^2 + R_y^2} = \sqrt{(49.5 \text{ units})^2 + (27.1 \text{ units})^2} = 56.4$ units

$\theta = \tan^{-1}\left(\dfrac{R_y}{R_x}\right) = \tan^{-1}\left(\dfrac{27.1}{49.5}\right) = 28.7°$

---

45. Consider two vectors $\mathbf{A} = 3\mathbf{i} - 2\mathbf{j}$ and $\mathbf{B} = -\mathbf{i} - 4\mathbf{j}$. Calculate (a) $\mathbf{A} + \mathbf{B}$, (b) $\mathbf{A} - \mathbf{B}$, (c) $|\mathbf{A} + \mathbf{B}|$, (d) $|\mathbf{A} - \mathbf{B}|$, (e) the directions of $\mathbf{A} + \mathbf{B}$ and $\mathbf{A} - \mathbf{B}$.

Solution   Use the property of vector addition that states that

If $\mathbf{R} = \mathbf{A} + \mathbf{B}$ then $R_x = A_x + B_x$ and $R_y = A_y + B_y$

(a) $\mathbf{A} + \mathbf{B} = (3\mathbf{i} - 2\mathbf{j}) + (-\mathbf{i} - 4\mathbf{j}) = 2\mathbf{i} - 6\mathbf{j}$

(b) $\mathbf{A} - \mathbf{B} = (3\mathbf{i} - 2\mathbf{j}) - (-\mathbf{i} - 4\mathbf{j}) = 4\mathbf{i} + 2\mathbf{j}$

For a vector, $\mathbf{R} = R_x\mathbf{i} + R_y\mathbf{j}$ $|\mathbf{R}| = \sqrt{R_x^2 + R_y^2}$

(c) $|\mathbf{A} + \mathbf{B}| = \sqrt{2^2 + (-6)^2} = 6.32$

(d) $|\mathbf{A} - \mathbf{B}| = \sqrt{4^2 + 2^2} = 4.47$

The direction of a vector relative to the positive x axis is $\theta = \tan^{-1}(R_y/R_x)$

(e) For $\mathbf{A} + \mathbf{B}$, $\theta = \tan^{-1}(-6/2) = -71.6° = 288°$

For $\mathbf{A} - \mathbf{B}$, $\theta = \tan^{-1}(2/4) = 26.6°$

---

49. The vector **A** has x, y, and z components of 8.00, 12.0, and –4.00 units, respectively. (a) Write a vector expression for **A** in unit-vector notation. (b) Obtain a unit-vector expression for a vector **B** one-fourth the length of **A** pointing in the same direction as **A**. (c) Obtain a unit-vector expression for a vector **C** three times the length of **A** pointing in the direction opposite the direction of **A**.

Solution

(a) $\mathbf{A} = A_x\mathbf{i} + A_y\mathbf{j} + A_z\mathbf{k}$ $\mathbf{A} = 8.00\,\mathbf{i} + 12.0\,\mathbf{j} - 4.00\,\mathbf{k}$ ◊

(b) $\mathbf{B} = \mathbf{A}/4$ $\mathbf{B} = 2.00\,\mathbf{i} + 3.00\,\mathbf{j} - 1.00\,\mathbf{k}$ ◊

(c) $\mathbf{C} = -3\mathbf{A}$ $\mathbf{C} = -24.0\,\mathbf{i} - 36.0\,\mathbf{j} + 12.0\,\mathbf{k}$ ◊

---

59. There are nearly $\pi \times 10^7$ s in one year. Find the percentage error in this approximation, where "percentage error" is defined as

$$\frac{\left|\text{assumed value} \pm \text{true value}\right|}{\text{true value}} \times 100\%$$

Solution   First evaluate the "true value." Remember that every fourth year is a leap year; therefore there are 365.25 days in an average year.

$$1\ \text{yr} = 1\ \text{yr}\left(\frac{365.25\ \text{d}}{1\ \text{yr}}\right)\left(\frac{24\ \text{h}}{1\ \text{d}}\right)\left(\frac{3600\ \text{s}}{1\ \text{h}}\right) = 3.1558 \times 10^7\ \text{s}$$

$$\text{Percentage error} = \frac{\left|\text{assumed value} - \text{true value}\right|}{\text{true value}} \times 100\%$$

$$\text{Percentage error} = \frac{(3.1558 - 3.1416) \times 10^7}{3.1558 \times 10^7} \times 100\%$$

$$\text{Percentage error} = 0.449\%$$ ◊

Chapter 2

Motion in One Dimension

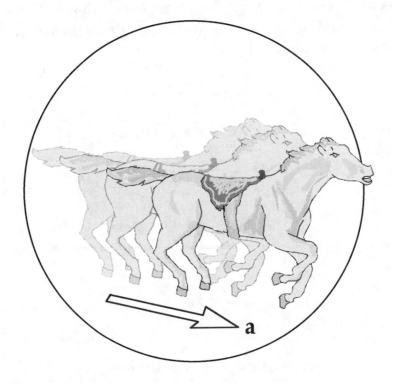

INTRODUCTION

As a first step in studying mechanics, it is convenient to describe motion in terms of space and time, ignoring for the present the agents that caused that motion. This portion of mechanics is called **kinematics.** In this chapter we consider motion along a straight line, that is, one-dimensional motion. Starting with the concept of displacement discussed in the previous chapter, we first define velocity and acceleration. Then, using these concepts, we study the motion of objects traveling in one dimension under a constant acceleration. In Chapter 3 we shall discuss the motions of objects in two dimensions.

In this chapter, we are concerned only with translational motion. In many situations, we can model the moving object as a particle, which in mathematics is defined as a point having no size.

NOTES FROM SELECTED CHAPTER SECTIONS

2.2 Instantaneous Velocity

The velocity of a particle at any instant of time (i.e. at some point on a space-time graph) is called the **instantaneous velocity**.

The **instantaneous speed** of an object, which is a scalar quantity, is defined as the magnitude of the instantaneous velocity. Hence, by definition, **speed can never be negative.**

The slope of the line tangent to the position-time curve at a point P is defined to be the **instantaneous velocity** at the corresponding time.

The area under the v vs. t curve in any time interval equals the displacement of the particle during that time interval.

2.4 Acceleration

The **average acceleration** during a given time interval is defined as the change in velocity divided by the time interval during which this change occurs.

The **instantaneous acceleration** of an object at a certain time equals the slope of the tangent to the velocity-time graph at that instant of time.

2.7 Freely Falling Objects

A freely falling body is an object moving freely under the influence of gravity only, regardless of its initial motion. Objects thrown upward or downward and those released from rest are all falling freely once they are released!

Once they are in free fall, all objects have an acceleration downward equal to the acceleration due to gravity. This is true regardless of the initial motion of the object.

EQUATIONS AND CONCEPTS

The **displacement** Δx of a particle moving from position x_i to position x_f equals the final coordinate minus the initial coordinate.

$$\Delta x \equiv x_f - x_i$$

Displacement should not be confused with distance traveled. When $x_f = x_i$, the displacement is zero; however, if the particle leaves x_i, travels along a path, and returns to x_i, the distance traveled will not be zero.

The x-component of the average velocity of a particle during a time interval is the ratio of the total displacement to the time interval during which the displacement occurred. Note that the instantaneous velocity of the object during the time interval might have different values from instant to instant.

$$\bar{v}_x = \frac{\Delta x}{\Delta t} = \frac{x_f - x_i}{t_f - t_i} \qquad (2.2)$$

The **average speed** of a particle is defined as the ratio of the total distance traveled to the time required to travel that distance.

The **instantaneous velocity** v_x is defined as the limit of the ratio $\Delta x / \Delta t$ as Δt approaches zero. The instantaneous velocity can be positive, negative, or zero.

$$v_x \equiv \lim_{\Delta t \to 0} \frac{\Delta x}{\Delta t} = \frac{dx}{dt} \qquad (2.3)$$

The **average acceleration** of an object during a time interval is the ratio of the change in velocity to the time interval during which the change in velocity occurs.

$$\bar{a}_x \equiv \frac{v_{xf} - v_{xi}}{t_f - t_i} = \frac{\Delta v_x}{\Delta t} \qquad (2.5)$$

The **instantaneous acceleration** a_x is defined as the limit of the ratio $\Delta v_x / \Delta t$ as Δt approaches zero.

$$a_x \equiv \lim_{\Delta t \to 0} \frac{\Delta v_x}{\Delta t} = \frac{dv_x}{dt} \qquad (2.6)$$

Equations 2.8 - 2.12 can be used to describe one-dimensional motion along the x axis of a **particle under constant acceleration**. Note that each equation shows a different relationship among physical quantities: initial velocity, final velocity, acceleration, time, and position.

$$v_{xf} = v_{xi} + a_x t \qquad (2.8)$$

$$\bar{v}_x = \frac{1}{2}\left(v_{xi} + v_{xf}\right) \qquad (2.9)$$

$$x_f = x_i + \frac{1}{2}(v_{xi} + v_{xf})t \qquad (2.10)$$

$$x_f = x_i + v_{xi} t + \frac{1}{2} a_x t^2 \qquad (2.11)$$

$$v_{xf}^2 = v_{xi}^2 + 2a_x\left(x_f - x_i\right) \qquad (2.12)$$

The only modifications we need to make in the above equations for freely falling objects is to note that the motion is vertical (along the y-axis), $+y$ is directed upward, and the acceleration is $a_y = -g$.

$$v_{yf} = v_{yi} - gt$$

$$y_f = y_i + \frac{1}{2}(v_{yf} + v_{yi})t$$

$$y_f = y_i + v_{yi}t - \frac{1}{2}gt^2$$

$$v_{yf}^2 = v_{yi}^2 - 2g(y_f - y_i)$$

SUGGESTIONS, SKILLS, AND STRATEGIES

The following procedure is recommended for solving problems that involve a particle under constant acceleration:

- Make sure all the units in the problem are consistent. That is, if distances are measured in meters, be sure that velocities have units of m/s and accelerations have units of m/s^2.

- Choose a coordinate system.

- Choose an instant to be called the "initial point," and another to be called the "final point."

- Think about what is going on physically in the problem, and then select from the list of kinematic equations those that will enable you to determine the unknowns.

- Construct an appropriate motion diagram or a graphical representation. Check to see if your answers are consistent with the diagram.

REVIEW CHECKLIST

▷ Define the displacement and average velocity of a particle in motion. Define the instantaneous velocity and understand how this quantity differs from average velocity.

▷ Define average acceleration and instantaneous acceleration.

▷ Construct a graph of position versus time (given a function such as $x = 5 + 3t - 2t^2$) for a particle in motion along a straight line. From this graph, you should be able to determine both average and instantaneous values of velocity by calculating the slope of the tangent to the graph.

▷ Describe what is meant by a body in **free fall** (one moving under the influence of gravity—where air resistance is neglected). Recognize that the equations of kinematics apply directly to a freely falling object and that the acceleration is then given by $a_y = -g$ (where $g = 9.80 \ m/s^2$).

▷ Apply the equations of kinematics to any situation where the motion occurs under constant acceleration.

ANSWERS TO SELECTED CONCEPTUAL QUESTIONS

4. Consider the following combinations of signs and values for velocity and acceleration of a particle with respect to a one-dimensional x axis:

	(a)	(b)	(c)	(d)	(e)	(f)	(g)	(h)
Velocity	+	+	+	−	−	−	0	0
Acceleration	+	−	0	+	−	0	+	−

Describe what a particle is doing in each case, and give a real-life example for an automobile on an east–west one-dimensional axis, with east considered the positive direction.

Answer

(a) The particle is moving to the right (in the +x direction) since the velocity is positive, and its speed is increasing, because the acceleration is in the same direction as the velocity. This would be the case in an automobile that is moving toward the east, starting up after waiting for a red light.

(b) The particle is moving to the right, since the velocity is positive, and its speed is decreasing, because the acceleration is in the opposite direction to the velocity. An automobile is moving toward the east, slowing down in preparation for a red light.

(c) The particle is moving to the right, since the velocity is positive, and its speed is constant, because the acceleration is zero. An automobile is moving toward the east, moving at constant speed on a freeway.

(d) The particle is moving to the left (in the −x direction), since the velocity is negative, and its speed is decreasing, because the acceleration is in the opposite direction to the velocity. An automobile is moving toward the west, slowing down in preparation for a red light.

(e) The particle is moving to the left (in the −x direction), since the velocity is negative, and its speed is increasing, because the acceleration is in the same direction as the velocity. An automobile is moving toward the west, starting up after waiting for a red light.

(f) The particle is moving to the left (in the $-x$ direction), since the velocity is negative, and its speed is constant, because the acceleration is zero. An automobile is moving toward the west, moving at constant speed on a freeway.

(g) The particle is momentarily at rest, and, in the next instant, will be moving to the right. This situation can only exist for an instant, since as soon as the particle begins moving, the velocity will no longer be zero. This is the situation for an automobile just before it begins to move to the east from a standstill at a red light.

(h) The particle is momentarily at rest, and, in the next instant, will be moving to the left. This situation can only exist for an instant, since as soon as the particle begins moving, the velocity will no longer be zero. This is the situation for an automobile just before it begins to move to the west from a standstill at a red light.

6. A student at the top of a building of height h throws one ball upward with an initial speed v_i and then throws a second ball downward with the same initial speed. How do the final velocities of the balls compare when they reach the ground?

Answer They are the same. After the first ball reaches its apex and falls back downward past the student, it will have a downward velocity equal to v_i. This velocity is the same as the velocity of the second ball, so after they fall through equal heights their impact speeds will also be the same.

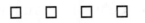

7. Two cars are moving in the same direction in parallel lanes along a highway. At some instant, the velocity of car A exceeds the velocity of car B. Does this mean that the acceleration of A is greater than that of B? Explain.

Answer No. If Car A has been traveling with cruise control, its velocity will be high (60 mph), but its acceleration will be close to zero. If Car B is pulling onto the highway, its velocity is likely to be low (30 mph), but its acceleration will be high.

SOLUTIONS TO SELECTED END-OF-CHAPTER PROBLEMS

3. The displacement versus time for a certain particle moving along the x axis is shown in Figure P2.3. Find the average velocity in the time intervals (a) 0 to 2 s, (b) 0 to 4 s, (c) 2 s to 4 s, (d) 4 s to 7 s, (e) 0 to 8 s.

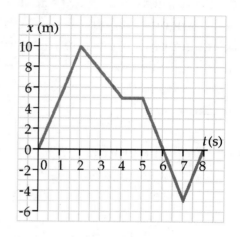

Figure P2.3

Solution

On this graph, we can tell positions to two significant figures:

(a) $x = 0$ at $t = 0$ and $x = 10$ m at $t = 2$ s

$$\overline{v}_x = \frac{\Delta x}{\Delta t} = \frac{10 \text{ m} - 0}{2 \text{ s} - 0} = 5.0 \text{ m / s}$$ ◊

(b) $x = 5.0$ m at $t = 4$ s

$$\overline{v}_x = \frac{\Delta x}{\Delta t} = \frac{5.0 \text{ m} - 0}{4 \text{ s} - 0} = 1.2 \text{ m / s}$$ ◊

(c) $\overline{v}_x = \dfrac{\Delta x}{\Delta t} = \dfrac{5.0 \text{ m} - 10 \text{ m}}{4 \text{ s} - 2 \text{ s}} = -2.5 \text{ m / s}$ ◊

(d) $\overline{v}_x = \dfrac{\Delta x}{\Delta t} = \dfrac{-5.0 \text{ m} - 5.0 \text{ m}}{7 \text{ s} - 4 \text{ s}} = -3.3 \text{ m / s}$ ◊

(e) $\overline{v}_x = \dfrac{\Delta x}{\Delta t} = \dfrac{0.0 \text{ m} - 0.0 \text{ m}}{8 \text{ s} - 0 \text{ s}} = 0 \text{ m / s}$ ◊

7. A position-time graph for a particle moving along the x axis is shown in Figure P2.7. (a) Find the average velocity in the time interval $t = 1.5$ s to $t = 4.0$ s. (b) Determine the instantaneous velocity at $t = 2.0$ s by measuring the slope of the tangent line shown in the graph. (c) At what value of t is the velocity zero?

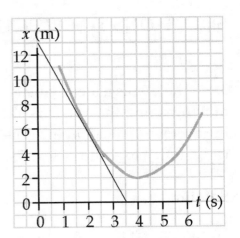

Solution

(a) From the graph:

At $t_1 = 1.5$ s, $x = x_1 = 8.0$ m

Figure P2.7

At $t_2 = 4.0$ s, $x = x_2 = 2.0$ m

Therefore, $\quad\quad\quad \bar{v}_{1\to 2} = \dfrac{\Delta x}{\Delta t} = \dfrac{2.0 \text{ m} - 8.0 \text{ m}}{4.0 \text{ s} - 1.5 \text{ s}} = -2.4 \text{ m} / \text{s}$ ◊

(b) Choose two points along line which is tangent to the curve at $t = 2.0$ s. We will use the two points:

$(t_i = 0.0 \text{ s}, \ x_i = 13.0 \text{ m})$ and $(t_f = 3.5 \text{ s}, \ x_f = 0.0 \text{ m})$

Instantaneous velocity equals the slope of the tangent line, so

$$v = \frac{x_f - x_i}{t_f - t_i} = \frac{0.0 \text{ m} - 13.0 \text{ m}}{3.5 \text{ s} - 0.0 \text{ s}} = -3.7 \text{ m/s}$$

The negative sign indicates that the **direction** of **v** is along the negative x direction. ◊

(c) The velocity will be zero when the slope of the tangent line is zero. This occurs for the point on the graph where x has its minimum value. Therefore,

$$v = 0 \ \text{ at } \ t \cong 4.0 \text{ s} \qquad\qquad ◊$$

15. A particle moves along the x axis according to the equation $x = 2.00 + 3.00t - t^2$, where x is in meters and t is in seconds. At $t = 3.00$ s, find (a) the position of the particle, (b) its velocity, and (c) its acceleration.

Solution With the position given by $x = 2.00 + 3.00t - t^2$, we can use the rules for differentiation to write expressions for the velocity and acceleration as functions of time:

$$v = \frac{dx}{dt} = 3.00 - 2t \qquad \text{and} \qquad a = \frac{dv}{dt} = -2$$

Now we can evaluate x, v, and a at $t = 3.00$ s.

(a) $x = 2.00 + 3.00(3.00) - (3.00)^2 = 2.00$ m   ◊

(b) $v = 3.00 - 2(3.00) = -3.00$ m/s   ◊

(c) $a = -2.00$ m/s²   ◊

21. An object moving with uniform acceleration has a velocity of 12.0 cm/s in the positive x direction when its x coordinate is 3.00 cm. If its x coordinate 2.00 s later is –5.00 cm, what is its acceleration?

Solution Take $t = 0$ to be the time when $x_i = 3.00$ cm and $v_{xi} = 12.0$ cm/s. Also, at $t = 2.00$ s, $x_f = -5.00$ cm. Use the kinematic equation $x_f = x_i + v_{xi}t + \frac{1}{2}at^2$, and solve for a.

$$a = \frac{2[x_f - x_i - v_{xi}t]}{t^2} \qquad \text{(acceleration is to the left)}$$

$$a = \frac{2[-5.00 \text{ cm} - 3.00 \text{ cm} - (12.0 \text{ cm/s})(2.00 \text{ s})]}{(2.00 \text{ s})^2}$$

$$a = -16.0 \text{ cm/s}^2 \qquad\qquad\qquad ◊$$

25. A jet plane comes in for landing with a speed of 100 m/s and can accelerate at a maximum rate of −5.00 m/s² as it comes to rest. (a) From the instant the plane touches the runway, what is the minimum time needed before it can come to rest? (b) Can this plane land at a small tropical island airport where the runway is 0.800 km long?

Solution

The negative acceleration of the plane as it lands is normally called **deceleration**; however, physicists tend to use the single term **acceleration** for both cases.

(a) Assume that the acceleration of the plane is **constant** at the maximum rate, so that the plane can be modeled as a particle under constant acceleration $a = -5.00$ m/s². Given $v_i = 100$ m/s and $v_f = 0$, use the equation $v_f = v_i + at$ and solve for t:

$$t = \frac{v_f - v_i}{a} = \frac{0 - 100 \text{ m/s}}{-5.00 \text{ m/s}^2} = 20.0 \text{ s} \qquad \Diamond$$

(b) Find the required stopping distance and compare this to the length of the runway. Taking x_i to be zero, we get

$$v_f^2 = v_i^2 + 2a(x_f - x_i)$$

or $\qquad \Delta x = x_f - x_i = \dfrac{v_f^2 - v_i^2}{2a} = \dfrac{0 - (100 \text{ m/s})^2}{2(-5.00 \text{ m/s}^2)} = 1000 \text{ m}$

The stopping distance is greater than the length of the runway; the plane **cannot land.** $\Diamond$

---

29. For many years, the world's land-speed record was held by Colonel John P. Stapp, USAF. On March 19, 1954, he rode a rocket-propelled sled that moved down a track at 632 mi/h. He and the sled were safely brought to rest in 1.40 s. Determine (a) the negative acceleration he experienced and (b) the distance he traveled during this negative acceleration.

Solution

We assume the acceleration remains constant during the 1.40 s period of negative acceleration.

$$v_i = 632 \, \frac{\text{mi}}{\text{h}} = 632 \, \frac{\text{mi}}{\text{h}} \left(\frac{1609 \text{ m}}{1 \text{ mi}} \right) \left(\frac{1 \text{ h}}{3600 \text{ s}} \right) = 282 \text{ m/s}$$

(a) Taking $v_f = v_i + at$ with $v_f = 0$, $a = \dfrac{v_f - v_i}{t} = \dfrac{0 - 282 \text{ m/s}}{1.40 \text{ s}} = -202 \text{ m/s}^2$ ◊

This is approximately $20g$!

(b) $x_f - x_i = \frac{1}{2}(v_i + v_f)t = \frac{1}{2}(282 \text{ m/s} + 0)(1.40 \text{ s}) = 198 \text{ m}$ ◊

---

35. A student throws a set of keys vertically upward to her sorority sister, who is in a window 4.00 m above. The keys are caught 1.50 s later by the sister's outstretched hand. (a) With what initial velocity were the keys thrown? (b) What was the velocity of the keys just before they were caught?

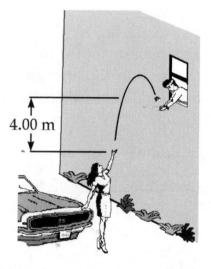

4.00 m

Solution   We model the keys as a particle under the constant free-fall acceleration. Taking the student's position to be $y_i = 0$, and given that $y = 4.00$ m at $t = 1.50$ s, we find (with $a = -9.80 \text{ m/s}^2$)

(a) $y_f = y_i + v_i t + \frac{1}{2}at^2$ or $v_i = \dfrac{y_f - y_i - \frac{1}{2}at^2}{t}$

$$v_i = \frac{4.00 \text{ m} - \frac{1}{2}(-9.80 \text{ m/s}^2)(1.50 \text{ s})^2}{1.50 \text{ s}} = 10.0 \text{ m/s}$$ ◊

(b) The velocity at any time $t > 0$ is given by $v_f = v_i + at$. Therefore, at $t = 1.50$ s,

$v_f = 10.0 \text{ m/s} - (9.80 \text{ m/s}^2)(1.50 \text{ s}) = -4.68 \text{ m/s}$

The negative sign means that the keys are moving **downward** just before they are caught.

---

37.   A baseball is hit so that it travels straight upward after being struck by the bat. A fan observes that it takes 3.00 s for the ball to reach its maximum height. Find (a) its initial velocity and (b) the height reached by the ball.

Solution   After leaving the bat, the ball is in free fall and has a constant acceleration, $a = -g = -9.80 \text{ m/s}^2$.

(a)   Solve the equation $v_f = v_i + at$ for v_i: $v_i = v_f - at$

When $t = 3.00$ s (at maximum height), $v_f = 0$

Therefore, $v_i = 0 - (-9.80 \text{ m/s}^2)(3.00 \text{ s}) = 29.4 \text{ m/s}$ ◊

(b)   When $t = 3.00$ s $y = v_i t + \frac{1}{2}at^2$ and $y = y_{max}$

So, $y_{max} = (29.4 \text{ m/s})(3.00 \text{ s}) + \frac{1}{2}\left(-9.80 \text{ m/s}^2\right)(3.00 \text{ s})^2 = 44.1 \text{ m}$ ◊

---

39.   A daring ranch hand sitting on a tree limb wishes to drop vertically onto a horse galloping under the tree. The constant speed of the horse is 10.0 m/s, and the distance from the limb to the level of the saddle is 3.00 m. (a) What must be the horizontal distance between the saddle and limb when the ranch hand makes his move? (b) How long is he in the air?

Solution   We do part (b) first.

(b)   Consider the vertical motion of the man after leaving the limb (with $v_i = 0$ at $y_i = 3.00$ m) until reaching the saddle (at $y_f = 0$). Modeling the man as a particle under constant acceleration, we find his time of fall from

$$y_f = y_i + v_i t + \frac{1}{2}at^2$$

When $v_i = 0$, $t = \sqrt{\dfrac{2(y_f - y_i)}{a}} = \sqrt{\dfrac{2(0 - 3.00 \text{ m})}{-9.80 \text{ m/s}^2}} = 0.782 \text{ s}$ ◊

(a)   During this time interval, the horse is modeled as a particle under constant velocity in the horizontal direction.

$v_i = v_f = 10.0 \text{ m/s}$ so $x_f - x_i = v_i t = (10.0 \text{ m/s})(0.782 \text{ s}) = 7.82 \text{ m}$

and the ranch hand must let go when the horse is 7.82 m from the tree.   ◊

---

47. Setting a new world record in a 100-m race, Maggie and Judy cross the finish line in a dead heat, both taking 10.2 s. Accelerating uniformly, Maggie took 2.00 s and Judy 3.00 s to attain maximum speed, which they maintained for the rest of the race. (a) What was the acceleration of each sprinter? (b) What were their respective maximum speeds? (c) Which sprinter was ahead at the 6.00-s mark, and by how much?

Solution

(a) Maggie moves with constant positive acceleration a_M for 2.00 s, then with constant speed (zero acceleration) for 8.20 s, covering a distance of $x_{M1} + x_{M2} = 100$ m.

$$x_{M1} = \tfrac{1}{2}a_M(2.00\text{ s})^2 \quad\text{and}\quad x_{M2} = v_M(8.20\text{ s}),$$

where v_M is her maximum speed. Since $v_M = 0 + a_M(2.00\text{ s})$,

by substitution $\qquad \tfrac{1}{2}a_M(2.00\text{ s})^2 + a_M(2.00\text{ s})(8.20\text{ s}) = 100$ m

$$a_M = 5.43 \text{ m} / \text{s}^2 \qquad\qquad \Diamond$$

Similarly, for Judy, $\quad x_{J1} + x_{J2} = 100$ m

with $\qquad x_{J1} = \tfrac{1}{2}a_J(3.00\text{ s})^2;\qquad x_{J2} = v_J(7.20\text{ s});\qquad v_J = a_J(3.00\text{ s})$

$$\tfrac{1}{2}a_J(3.00\text{ s})^2 + a_J(3.00\text{ s})(7.20\text{ s}) = 100\text{ m}$$

$$a_J = 3.83 \text{ m}/\text{s}^2 \qquad\qquad \Diamond$$

(b) Their speeds after accelerating are

$$v_M = a_M(2.00\text{ s}) = (5.43\text{ m}/\text{s}^2)(2.00\text{ s}) = 10.9\text{ m}/\text{s} \qquad \Diamond$$

and $\qquad v_J = a_J(3.00\text{ s}) = (3.83\text{ m}/\text{s}^2)(3.00\text{ s}) = 11.5\text{ m}/\text{s} \qquad \Diamond$

(c) In the first 6.00 s, Maggie covers a distance

$$\tfrac{1}{2}a_M(2.00\text{ s})^2 + v_M(4.00\text{ s}) = \tfrac{1}{2}\left(5.43\text{ m}/\text{s}^2\right)(2.00\text{ s})^2 + (10.9\text{ m}/\text{s})(4.00\text{ s}) = 54.3\text{ m}$$

and Judy has run a distance

$$\tfrac{1}{2}a_J(3.00\text{ s})^2 + v_J(3.00\text{ s}) = \tfrac{1}{2}\left(3.83\text{ m}/\text{s}^2\right)(3.00\text{ s})^2 + (11.5\text{ m}/\text{s})(3.00\text{ s}) = 51.7\text{ m}$$

So Maggie is ahead by 54.3 m – 51.7 m = 2.62 m $\qquad\qquad \Diamond$

51. An inquisitive physics student and mountain climber climbs a 50.0-m cliff that overhangs a calm pool of water. He throws two stones vertically downward, 1.00 s apart, and observes that they cause a single splash. The first stone has an initial speed of 2.00 m/s. (a) How long after release of the first stone do the two stones hit the water? (b) What was the initial velocity of the second stone? (c) What is the velocity of each stone at the instant the two hit the water?

Solution   Set $y_i = 0$ at the top of the cliff, and find the time required for the first stone to reach the water using the particle under constant acceleration model:

$$y_f = y_i + v_i t + \tfrac{1}{2}at^2 \quad \text{or in quadratic form,} \quad -\tfrac{1}{2}at^2 - v_i t + \left(y_f - y_i\right) = 0$$

(a)   If we take the direction downward to be negative, $y_f = -50.0$ m, $v_i = -2.00$ m/s, and $a = -9.80$ m/s². Substituting these values into the equation, we find

$$\left(4.90 \text{ m}/\text{s}^2\right)t^2 + (2.00 \text{ m}/\text{s})t - 50.0 \text{ m} = 0$$

Using the quadratic formula, and noting that only the positive root describes this physical situation,

$$t = \frac{-2.00 \text{ m}/\text{s} \pm \sqrt{(2.00 \text{ m}/\text{s})^2 - 4(4.90 \text{ m}/\text{s}^2)(-50.0 \text{ m})}}{2(4.90 \text{ m}/\text{s}^2)} = 3.00 \text{ s} \qquad \Diamond$$

(b)   For the second stone, the time of travel is $t = 3.00 \text{ s} - 1.00 \text{ s} = 2.00 \text{ s}$

Since $y_f = y_i + v_i t + \tfrac{1}{2}at^2$,

$$v_i = \frac{(y_f - y_i) - \tfrac{1}{2}at^2}{t} = \frac{-50.0 \text{ m} - \left(\tfrac{1}{2}\right)\left(-9.80 \text{ m}/\text{s}^2\right)(2.00 \text{ s})^2}{2.00 \text{ s}} = -15.3 \text{ m}/\text{s} \qquad \Diamond$$

The negative value indicates the downward direction of the initial velocity of the second stone.

(c)   For the first stone, $v_{1f} = v_{1i} + a_1 t_1 = -2.00 \text{ m}/\text{s} + (-9.80 \text{ m}/\text{s}^2)(3.00 \text{ s})$

$$v_{1f} = -31.4 \text{ m}/\text{s} \qquad \Diamond$$

For the second stone, $v_{2f} = v_{2i} + a_2 t_2 = -15.3 \text{ m}/\text{s} + (-9.80 \text{ m}/\text{s}^2)(2.00 \text{ s})$

$$v_{2f} = -34.8 \text{ m}/\text{s} \qquad \Diamond$$

57. Two objects, A and B, are connected by a rigid rod that has a length L. The objects slide along perpendicular guide rails, as shown in Figure P2.57. If A slides to the left with a constant speed v, find the speed of B when $\alpha = 60.0°$.

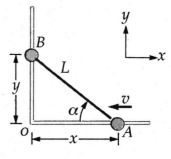

Solution We translate from a pictorial representation through a geometric model to a mathematical representation by observing that the distances x and y are always related by $x^2 + y^2 = L^2$. Differentiating this equation with respect to time, we have

Figure P2.57

$$2x\,\frac{dx}{dt} + 2y\,\frac{dy}{dt} = 0$$

Now $\dfrac{dy}{dt}$ is v_B, the unknown velocity of B; and $\dfrac{dx}{dt} = -v$.

So the differentiated equation becomes $\quad \dfrac{dy}{dt} = -\dfrac{x}{y}\left(\dfrac{dx}{dt}\right) = \left(-\dfrac{x}{y}\right)(-v) = v_B$

But $\qquad\qquad\qquad\qquad \dfrac{y}{x} = \tan\alpha, \quad \text{so} \quad v_B = \left(\dfrac{1}{\tan\alpha}\right)v$

When $\alpha = 60°$, $\qquad\qquad\qquad v_B = \dfrac{v}{\tan 60.0°} = \dfrac{v}{\sqrt{3}} = 0.577v$ ◊

Chapter 3

Motion in Two Dimensions

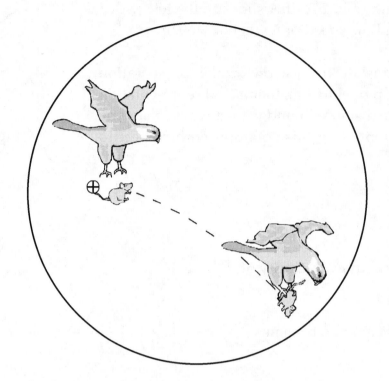

INTRODUCTION

In this chapter we deal with the kinematics of a particle moving in a plane, which is two-dimensional motion. Common examples of motion in a plane include the motion of projectiles and satellites in a gravitational field, and the motion of charged particles in uniform electric fields.

As in the case of one-dimensional motion, we derive the kinematic equations for two-dimensional motion from the fundamental definitions of displacement, velocity, and acceleration. As special cases of motion in two dimensions, we then treat constant-acceleration motion in a plane (projectile motion) and uniform circular motion.

NOTES FROM SELECTED CHAPTER SECTIONS

3.1 The Position, Velocity, and Acceleration Vectors

A particle's **position vector r** leads from the origin of a coordinate system to the instantaneous location of the particle. The **displacement vector Δr** is the final position vector minus the initial position vector. Note that the magnitude of the displacement vector is in general less than the distance measured along the actual path of travel.

It is important to recognize that a particle experiences an **acceleration** when:
- the magnitude of the velocity (speed) changes while the direction remains constant (e.g., a sphere rolling down an inclined plane);
- the velocity's magnitude remains constant while the velocity's direction changes (e.g., a particle moving at constant speed around a circle of constant radius);
- both the magnitude and direction of the velocity change (e.g., a mass vibrating up and down at the end of a spring).

3.3 Projectile Motion

Projectile motion is a common example of motion in two dimensions under constant acceleration. In the simplification model in which air resistance is negligible, the characteristics of projectile motion can be summarized as follows:
- The horizontal component of velocity, v_x, remains constant since the horizontal component of acceleration is zero.
- The vertical component of acceleration is downward and equal to the acceleration due to gravity, g.
- The vertical component of velocity, v_y, and the position in the y direction change in time in a manner identical to that of a freely-falling body.
- Projectile motion can be described as a superposition, or vector addition, of the two motions in the x and y directions.

The trajectory of a projectile is a **parabola** as shown in Figure 3.1. We choose the motion to be in the x-y plane, and take the **initial** velocity of the projectile to have a magnitude of v_i, directed at an angle θ_i with the horizontal. The parabolic path of travel is completely determined when the magnitude, v_i, and the direction, θ_i, of the initial velocity vector are given. The actual motion of the projectile is a superposition of two motions: motion of a freely-falling body with constant acceleration in the vertical direction, $-g$, and motion in the horizontal direction with constant velocity, $v_x = v_i \cos\theta_i$.

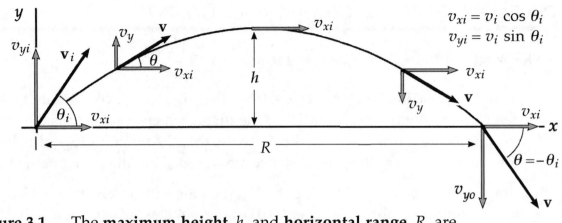

$$v_{xi} = v_i \cos \theta_i$$
$$v_{yi} = v_i \sin \theta_i$$

Figure 3.1 The **maximum height** h and **horizontal range** R are special coordinates defined here. They can be obtained from Equations 3.15 and 3.16.

3.4 The Particle in Uniform Circular Motion

The **centripetal** acceleration is the acceleration experienced by a mass which moves uniformly in a circular path of constant radius. The **direction** of the centripetal acceleration is always toward the center of the circular path.

Uniform circular motion is the motion of an object moving in a circular path with **constant linear** speed. The velocity vector is always tangent to the path of the moving body and therefore **perpendicular** to the radius.

3.5 Tangential and Radial Acceleration

Tangential acceleration of a particle moving in a circular path is due to a change in the speed (magnitude of the velocity vector) of the particle. The direction of the tangential acceleration at any instant is tangent to the circular path (perpendicular to the radius).

If a particle moves in a circle such that the speed v is **not** constant, the components of acceleration and the total acceleration at some instant are as shown in Figure 3.2.

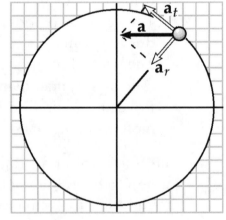

Figure 3.2
Components of acceleration when $|\mathbf{v}|$ is not constant

EQUATIONS AND CONCEPTS

A particle whose position vector changes from $\mathbf{r}_i$ to $\mathbf{r}_f$ undergoes a **displacement** $\Delta \mathbf{r} \equiv \mathbf{r}_f - \mathbf{r}_i$.

$$\Delta \mathbf{r} \equiv \mathbf{r}_f - \mathbf{r}_i \tag{3.1}$$

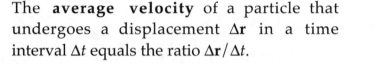

Figure 3.3

The **average velocity** of a particle that undergoes a displacement $\Delta \mathbf{r}$ in a time interval Δt equals the ratio $\Delta \mathbf{r}/\Delta t$.

$$\overline{\mathbf{v}} \equiv \frac{\Delta \mathbf{r}}{\Delta t} \tag{3.2}$$

The **instantaneous velocity** of a particle equals the limit of the average velocity as $\Delta t \to 0$.

$$\mathbf{v} \equiv \lim_{\Delta t \to 0} \frac{\Delta \mathbf{r}}{\Delta t} = \frac{d\mathbf{r}}{dt} \tag{3.3}$$

The **average acceleration** of a particle which undergoes a change in velocity $\Delta \mathbf{v}$ in a time interval Δt equals the ratio $\Delta \mathbf{v}/\Delta t$.

$$\overline{\mathbf{a}} \equiv \frac{\mathbf{v}_f - \mathbf{v}_i}{t_f - t_i} = \frac{\Delta \mathbf{v}}{\Delta t} \tag{3.4}$$

The **instantaneous acceleration** is defined as the limit of the average acceleration as $\Delta t \to 0$.

$$\mathbf{a} \equiv \lim_{\Delta t \to 0} \frac{\Delta \mathbf{v}}{\Delta t} = \frac{d\mathbf{v}}{dt} \tag{3.5}$$

The position vector for a particle moving in the x-y plane can be written in terms of the coordinates, which change with time.

$$\mathbf{r} = x\mathbf{i} + y\mathbf{j} \tag{3.6}$$

The **velocity** of a particle as a function of time moving with **constant acceleration**: (at $t = 0$, the velocity is $\mathbf{v}_i$).

$$\mathbf{v}_f = \mathbf{v}_i + \mathbf{a}\,t \tag{3.8}$$

The **position vector** as a function of time for a particle moving with **constant acceleration**: (at $t = 0$, the position vector is $\mathbf{r}_i$ and the velocity is $\mathbf{v}_i$).

$$\mathbf{r}_f = \mathbf{r}_i + \mathbf{v}_i t + \tfrac{1}{2}\mathbf{a}t^2 \tag{3.9}$$

Equations 3.10 and 3.11 give the x and y components of velocity versus time for a projectile.

$$v_{xf} = v_{xi} = v_i \cos\theta_i = \text{constant} \tag{3.10}$$

$$v_{yf} = v_{yi} - gt = v_i \sin\theta_i - gt \tag{3.11}$$

Equations 3.12 and 3.13 give the x and y position coordinates for a projectile, as a function of time.

$$x_f = (v_i \cos\theta_i)t \tag{3.12}$$

$$y_f = (v_i \sin\theta_i)t - \tfrac{1}{2}gt^2 \tag{3.13}$$

The trajectory of a projectile is a parabola. The path of the projectile is completely specified if the values of v_i and θ_i are known. Equation 3.14 is valid for angles in the range $0 < \theta_i < \pi/2$.

$$y_f = (\tan\theta_i)x_f - \left(\frac{g}{2v_i^2 \cos^2\theta_i}\right)x_f^2 \tag{3.14}$$

The **maximum height** of a projectile can be written in terms of v_i and θ_i.

$$h = \frac{v_i^2 \sin^2\theta_i}{2g} \tag{3.15}$$

Likewise, the **horizontal range** of a projectile can also be stated in terms of v_i and θ_i.

$$R = \frac{v_i^2 \sin 2\theta_i}{g} \tag{3.16}$$

A particle moving in a circle of radius r with speed v undergoes a **centripetal acceleration** a_c equal in magnitude to v^2/r.

$$a_c = \frac{v^2}{r} \tag{3.17}$$

If a particle moves with constant speed v in a circular path, the period is equal to the total distance traveled in one revolution, divided by the speed.

$$T \equiv \frac{2\pi r}{v} \tag{3.18}$$

In general, a particle moving on a curved path has both a radial component vector and tangential component vector of acceleration. The tangential component vector $\mathbf{a}_t$ is tangent to the path, and is directed forward if the speed is increasing. The radial component vector is directed inward toward the center of curvature of the path. Because radially outward from a center is defined as positive by convention, the radial component of acceleration is negative and equal in magnitude to the centripetal acceleration.

$$\mathbf{a} = \mathbf{a}_r + \mathbf{a}_t \tag{3.19}$$

$$a_r = -a_c$$

If the speed of a particle moving on a curved path changes with time, the particle has a **tangential** component of acceleration equal in magnitude to $d|\mathbf{v}|/dt$. The magnitude of the radial component of acceleration is given by Equation 3.17.

$$a_t = \frac{d|\mathbf{v}|}{dt} \tag{3.20}$$

SUGGESTIONS, SKILLS, AND STRATEGIES

- You should be familiar with the mathematical expression for a **parabola.** In particular, the equation which describes the trajectory of a projectile moving under the influence of gravity is given by

$$y = Ax - Bx^2$$

where $\qquad A = \tan \theta_i \qquad$ and $\qquad B = \dfrac{g}{2v_i^2 \cos^2 \theta_i}$

Note that this expression for y assumes that the particle leaves the origin at $t = 0$, with a velocity $\mathbf{v}_i$. A sketch of y versus x for this situation is shown in Figure 3.4.

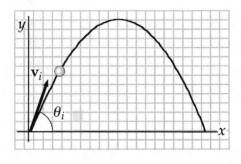

Figure 3.4

- If you are given v_i and θ_i, you should be able to make a point-by-point plot of the trajectory using the expressions for $x(t)$ and $y(t)$. Furthermore, you should know how to calculate the velocity component v_y at any time t. (Note that the component $v_x = v_{xi} = v_i \cos \theta_i = \text{constant}$, since $a_x = 0$.)

- Assuming that you have values for x and y at any time $t > 0$, you should be able to write an expression for the position vector $\mathbf{r}$ at that time using the relation $\mathbf{r} = x\mathbf{i} + y\mathbf{j}$. Likewise, if v_x and v_y are known at any time $t > 0$, you can express the velocity vector $\mathbf{v}$ in the formula $\mathbf{v} = v_x\mathbf{i} + v_y\mathbf{j}$. From this, you can find the **speed** at any time, since $v = \sqrt{v_x{}^2 + v_y{}^2}$.

Problem-Solving Strategy: Projectile Motion

You should use the following approach to solving projectile motion problems:

- Select a coordinate system.

- Resolve the initial velocity vector into x and y components.

- Treat the horizontal motion and the vertical motion independently.

- Follow the techniques for solving problems with constant velocity to analyze the horizontal motion of the projectile.

- Follow the techniques for solving problems with constant acceleration to analyze the vertical motion of the projectile.

REVIEW CHECKLIST

▷ Describe the position, velocity, and acceleration of a particle moving in the x-y plane.

▷ Recognize that two-dimensional motion in the x-y plane with constant acceleration is equivalent to two independent motions along the x and y directions with constant acceleration components a_x and a_y.

▷ Recognize the fact that if the initial speed v_i and initial angle θ_i of a projectile are known at a given point at $t = 0$, the velocity components and coordinates can be found at any later time t. Furthermore, one can also calculate the horizontal range R and maximum height h if v_i and θ_i are known.

▷ Understand the nature of the acceleration of a particle moving in a circle with constant speed. In this situation, note that although $|\mathbf{v}|$ = constant, the **direction** of **v** varies in time, the result of which is the radial, or centripetal acceleration.

▷ Describe the components of acceleration for a particle moving on a curved path, where both the magnitude and direction of **v** are changing with time. In this case, the particle has a tangential component of acceleration and a radial component of acceleration.

▷ Realize that the outcome of a measurement of the motion of a particle (its position, velocity, and acceleration) depends on the frame of reference of the observer.

ANSWERS TO SELECTED CONCEPTUAL QUESTIONS

1. If you know the position vectors of a particle at two points along its path and also know the time it took to move from one point to the other, can you determine the particle's instantaneous velocity? Its average velocity? Explain.

Answer Its instantaneous velocity cannot be determined at any point from this information. However, the average velocity over the time interval can be determined from its definition and the given information.

2. Explain whether or not the following particles have an acceleration: (a) a particle moving in a straight line with constant speed and (b) a particle moving around a curve with constant speed.

Answer (a) The acceleration is zero, since the magnitude and direction of **v** remain constant. (b) The particle has an acceleration since the direction of **v** changes.

□   □   □   □

4. A spacecraft drifts through space at a constant velocity. Suddenly a gas leak in the side of the spacecraft causes a constant acceleration of the spacecraft in a direction perpendicular to the initial velocity. The orientation of the spacecraft does not change, so that the acceleration remains perpendicular to the original direction of the velocity. What is the shape of the path followed by the spacecraft in this situation?

Answer The spacecraft will follow a parabolic path. This is equivalent to a projectile thrown off a cliff with a horizontal velocity. For the projectile, gravity provides an acceleration which is always perpendicular to the initial velocity, resulting in a parabolic path. For the spacecraft, the initial velocity plays the role of the horizontal velocity of the projectile. The leaking gas provides an acceleration that plays the role of gravity for the projectile. If the orientation of the spacecraft were to change in response to the gas leak (which is by far the more likely result), then the acceleration would change direction and the motion could become quite complicated.

□   □   □   □

9. A projectile is launched at some angle to the horizontal with some initial speed v_i, and air resistance is ignored. Is the projectile a freely falling body? What is its acceleration in the vertical direction? What is its acceleration in the horizontal direction?

Answer Yes. The acceleration is a freely falling body, because nothing counteracts the force of gravity. The vertical acceleration will be the local gravitational acceleration, g; the horizontal acceleration will be zero.

SOLUTIONS TO SELECTED END-OF-CHAPTER PROBLEMS

1. A motorist drives south at 20.0 m/s for 3.00 min, then turns west and travels at 25.0 m/s for 2.00 min, and finally travels northwest at 30.0 m/s for 1.00 min. For this 6.00-min trip, find (a) the total vector displacement, (b) the average speed, and (c) the average velocity. Let the positive x-axis point east.

Solution

(a) For each segment of the motion we model the car as a particle under constant velocity. Her displacements are

$$\Delta\mathbf{r} = (20.0 \text{ m}/\text{s})(180 \text{ s}) \textbf{ south} + (25.0 \text{ m}/\text{s})(120 \text{ s}) \textbf{ west} + (30.0 \text{ m}/\text{s})(60.0 \text{ s}) \textbf{ northwest}$$

Choosing $\mathbf{i}$ = east and $\mathbf{j}$ = north, we have

$$\Delta\mathbf{r} = (3.60 \text{ km})(-\mathbf{j}) + (3.00 \text{ km})(-\mathbf{i}) + (1.80 \text{ km } \cos 45.0°)(-\mathbf{i}) + (1.80 \text{ km } \sin 45.0°)(\mathbf{j})$$

$$\Delta\mathbf{r} = (3.00 + 1.27) \text{ km } (-\mathbf{i}) + (1.27 - 3.60) \text{ km } \mathbf{j} = (-4.27\mathbf{i} - 2.33\mathbf{j}) \text{ km} \qquad \lozenge$$

The answer can also be written as

$$\Delta\mathbf{r} = \sqrt{(-4.27 \text{ km})^2 + (-2.33 \text{ km})^2} \quad \text{at } \tan^{-1}\left(\frac{2.33}{4.27}\right) = 29° \text{ south of west,}$$

or $\qquad \Delta\mathbf{r} = 4.87$ km at 209° from east $\qquad \lozenge$

(b) The total distance or path-length traveled is $\qquad (3.60 + 3.00 + 1.80) \text{ km} = 8.40 \text{ km}$

$$\text{So average speed} = \frac{8.40 \text{ km}}{6.00 \text{ min}}\left(\frac{1.00 \text{ min}}{60.0 \text{ s}}\right)\left(\frac{1000 \text{ m}}{\text{km}}\right) = 23.3 \text{ m}/\text{s} \qquad \lozenge$$

(c) $\quad \bar{\mathbf{v}} = \dfrac{\Delta\mathbf{r}}{t} = \dfrac{4.87 \text{ km}}{360 \text{ s}} = 13.5 \text{ m}/\text{s at } 209° \qquad \lozenge$

or $\qquad \bar{\mathbf{v}} = \dfrac{\Delta\mathbf{r}}{t} = \dfrac{(-4.27 \textbf{ east} - 2.33 \textbf{ north}) \text{ km}}{360 \text{ s}} = (11.9 \textbf{ west} + 6.47 \textbf{ south}) \text{ m}/\text{s} \qquad \lozenge$

5. A fish swimming in a horizontal plane has velocity $\mathbf{v}_i = (4.00\mathbf{i} + 1.00\mathbf{j})$ m/s at a point in the ocean where the displacement from a certain rock is $\mathbf{r}_i = (10.0\mathbf{i} - 4.00\mathbf{j})$ m. After the fish swims with constant acceleration for 20.0 s, its velocity is $\mathbf{v}_f = (20.0\mathbf{i} - 5.00\mathbf{j})$ m/s. (a) What are the components of the acceleration? (b) What is the direction of the acceleration with respect to unit vector $\mathbf{i}$? (c) If the fish maintains constant acceleration where is the fish at $t = 25.0$ s and in what direction is it moving?

Solution Model the fish as a particle under constant acceleration.

At $t = 0$, $\mathbf{v}_i = (4.00\mathbf{i} + 1.00\mathbf{j})$ m/s and $\mathbf{r}_i = (10.0\mathbf{i} - 4.00\mathbf{j})$ m

At $t = 20.0$ s, $\mathbf{v}_f = (20.0\mathbf{i} - 5.00\mathbf{j})$ m/s

(a) $a_x = \dfrac{\Delta v_x}{\Delta t} = \dfrac{20.0 \text{ m/s} - 4.00 \text{ m/s}}{20.0 \text{ s}} = 0.800 \text{ m/s}^2$ ◊

$a_y = \dfrac{\Delta v_y}{\Delta t} = \dfrac{-5.00 \text{ m/s} - 1.00 \text{ m/s}}{20.0 \text{ s}} = -0.300 \text{ m/s}^2$ ◊

(b) $\theta = \tan^{-1}\left(\dfrac{a_y}{a_x}\right) = \tan^{-1}\left(\dfrac{-0.300 \text{ m/s}^2}{0.800 \text{ m/s}^2}\right) = -20.6°$ or 339° from the +x axis ◊

(c) At $t = 25.0$ s, its coordinates are

$x_f = x_i + v_{xi}t + \frac{1}{2}a_x t^2 = 10.0 \text{ m} + (4.00 \text{ m/s})(25.0 \text{ s}) + \frac{1}{2}(0.800 \text{ m/s}^2)(25.0 \text{ s})^2 = 360 \text{ m}$

$y_f = y_i + v_{yi}t + \frac{1}{2}a_y t^2 = -4.00 \text{ m} + (1.00 \text{ m/s})(25.0 \text{ s}) + \dfrac{(-0.300 \text{ m/s}^2)(25.0 \text{ s})^2}{2} = -72.8 \text{ m}$

$v_{yf} = v_{yi} + a_y t = (1.00 \text{ m/s}) - (0.300 \text{ m/s}^2)(25.0 \text{ s}) = -6.50 \text{ m/s}$

$v_{xf} = v_{xi} + a_x t = (4.00 \text{ m/s}) + (0.800 \text{ m/s}^2)(25.0 \text{ s}) = 24.0 \text{ m/s}$

Therefore, $\mathbf{r} = 360\mathbf{i} - 72.8\mathbf{j}$ m ◊

$\theta = \tan^{-1}\left(\dfrac{v_y}{v_x}\right) = \tan^{-1}\left(\dfrac{-6.50 \text{ m/s}}{24.0 \text{ m/s}}\right) = -15.2° = 345°$ from the +x axis. ◊

9. In a local bar, a customer slides an empty beer mug down the counter for a refill. The bartender is momentarily distracted and does not see the mug, which slides off the counter and strikes the floor 1.40 m from the base of the counter. If the height of the counter is 0.860 m, (a) with what velocity did the mug leave the counter, and (b) what was the direction of the mug's velocity just before it hit the floor?

Solution Choose the origin as the point where the mug just leaves the counter. Here, its motion is horizontal, with $v_{yi} = 0$. Choose the final point just before it hits the floor. Between these two points we model the mug as a particle moving with constant horizontal velocity and with constant vertical acceleration.

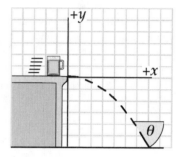

Here, $x_f = 1.40$ m, and $y_f = -0.860$ m; v_{xf} and v_{yf} are both unknown.

For the horizontal motion: $x_f = 1.40$ m, $a_x = 0$

For the vertical motion: $y_f = -0.860$ m, $v_{yi} = 0$, $a_y = -9.80$ m/s²

(a) To find v_{xf} using $x = v_{xf}t$, we need first to know the time of flight. Since it is the one quantity in common between horizontal and vertical motions, we find it from

$$y_f = v_{yi}t + \tfrac{1}{2}a_y t^2$$

Solving, $-0.860 \text{ m} = 0 + \tfrac{1}{2}(-9.80 \text{ m/s}^2)t^2$

and $t = 0.419$ s

Then, $v_{xf} = \dfrac{x_f}{t} = \dfrac{1.40 \text{ m}}{0.419 \text{ s}} = 3.34 \text{ m/s}$ ◊

(b) As the mug hits the floor, $v_{yf} = v_{yi} + a_y t = 0 - (9.80 \text{ m/s}^2)(0.419 \text{ s}) = -4.11 \text{ m/s}$

The impact angle, $\theta = \tan^{-1}\left(\dfrac{v_y}{v_x}\right) = \tan^{-1}\dfrac{-4.11 \text{ m / s}}{3.34 \text{ m / s}} = -50.9°$ ◊

15. A place kicker must kick a football from a point 36.0 m (about 40 yards) from the goal, and half the crowd hopes the ball will clear the crossbar, which is 3.05 m high. When kicked, the ball leaves the ground with a speed of 20.0 m/s at an angle of 53.0° to the horizontal. (a) By how much does the ball clear or fall short of clearing the crossbar? (b) Does the ball approach the crossbar while still rising or while falling?

Solution   Model the football as a projectile, moving with constant horizontal velocity and with constant vertical acceleration.

(a) To find the actual height of the football when it reaches the goal line, use Eq. 3.14:

$$y_f = x_f \tan \theta_i - \frac{g x_f^2}{2 v_i^2 \cos^2 \theta_i}$$

where $x_f = 36.0$ m, $v_i = 20.0$ m/s,   and $\theta_i = 53.0°$

So, $y_f = (36.0 \text{ m})(\tan 53.0°) - \dfrac{(9.80 \text{ m/s}^2)(36.0 \text{ m})^2}{(2)(20.0 \text{ m/s})^2 \cos^2 53.0°} = 47.774 - 43.834 = 3.939$ m

The ball clears the bar by $(3.939 - 3.050)$ m $= 0.889$ m.   ◊

(b) The time the ball takes to reach the maximum height ($v_y = 0$) is

$$t_1 = \frac{\left(v_i \sin \theta_i\right) - v_y}{g} = \frac{(20.0 \text{ m/s})(\sin 53.0°) - 0}{9.80 \text{ m/s}^2} = 1.63 \text{ s}$$

The time to travel 36.0 m horizontally is $t_2 = \dfrac{x_f}{v_{xi}}$

$$t_2 = \frac{36.0 \text{ m}}{(20.0 \text{ m / s})(\cos 53.0°)} = 2.99 \text{ s}$$

Since $t_2 > t_1$, the ball clears the goal on its way down.   ◊

25. The athlete shown in Figure P3.25 of the textbook rotates a 1.00-kg discus along a circular path of radius 1.06 m. The maximum speed of the discus is 20.0 m/s. Determine the magnitude of the maximum radial acceleration of the discus.

Solution   The maximum radial acceleration occurs when maximum tangential speed is attained. Model the discus here as a particle in uniform circular motion.

Here, $$a_c = \frac{v^2}{r} = \frac{(20.0 \text{ m/s})^2}{(1.06 \text{ m})} = 377 \text{ m/s}^2 \qquad \Diamond$$

29. A train slows down as it rounds a sharp horizontal turn, slowing from 90.0 km/h to 50.0 km/h in the 15.0 s that it takes to round the bend. The radius of the curve is 150 m. Compute the acceleration at the moment the train speed reaches 50.0 km/h, assuming that it continues to slow down at this time at the same rate.

Solution

$$50.0 \text{ km/h} = \left(50.0 \frac{\text{km}}{\text{h}}\right)\left(10^3 \frac{\text{m}}{\text{km}}\right)\left(\frac{1 \text{ h}}{3600 \text{ s}}\right) = 13.89 \text{ m/s}$$

$$90.0 \text{ km/h} = \left(90.0 \frac{\text{km}}{\text{h}}\right)\left(10^3 \frac{\text{m}}{\text{km}}\right)\left(\frac{1 \text{ h}}{3600 \text{ s}}\right) = 25.0 \text{ m/s}$$

Therefore, when $v = 13.89$ m/s,

$$a_r = -\frac{v^2}{r} = -\frac{(13.89 \text{ m/s})^2}{150 \text{ m}} = -1.29 \text{ m/s}^2 \text{ (inward)} \qquad a_t = \frac{\Delta v}{\Delta t} = \frac{13.89 \text{ m/s} - 25.0 \text{ m/s}}{15.0 \text{ s}} = -0.741 \text{ m/s}^2$$

The magnitude of the total acceleration is

$$a = \sqrt{a_r^2 + a_t^2} = \sqrt{(-1.29 \text{ m/s}^2)^2 + (-0.741 \text{ m/s}^2)^2} = 1.48 \text{ m/s}^2 \qquad \Diamond$$

Its direction is inward and backward, making an angle of

$$\theta = \tan^{-1}(a_t/a_r) = \tan^{-1}(0.741/1.29) = 29.9° \text{ behind the radial line.} \qquad \Diamond$$

31. Figure P3.31 represents the total acceleration of a particle moving clockwise in a circle of radius 2.50 m at a certain time. At this instant, find (a) the radial acceleration, (b) the speed of the particle, and (c) its tangential acceleration.

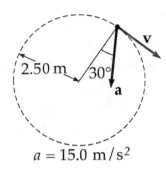

$a = 15.0 \text{ m/s}^2$

Solution

Figure P3.31

(a) The acceleration has an inward radial component:

$$a_c = a\cos 30.0° = \left(15.0 \text{ m/s}^2\right)\cos 30.0° = 13.0 \text{ m/s}^2$$

◊

The speed at the instant shown can be found by using

$$a_c = \frac{v^2}{r} \qquad \text{or} \qquad v = \sqrt{a_c r} = \sqrt{(13.0 \text{ m/s}^2)(2.50 \text{ m})} = 5.70 \text{ m/s}$$

◊

(c) The acceleration also has a tangential component:

$$a_t = a \ \sin \ 30.0° = (15.0 \text{ m/s}^2) \ \sin \ 30.0° = 7.50 \text{ m/s}^2$$

◊

---

33. A river has a steady speed of 0.500 m/s. A student swims upstream a distance of 1.00 km and swims back to the starting point. If the student can swim at a speed of 1.20 m/s in still water, how long does the trip take? Compare this with the time the trip would take if the water were still.

Solution

If we think about the time for the trip as a function of the stream's speed, we realize that if the stream is flowing at the same rate or faster than the student can swim, he will never reach the 1.00 km mark even after an infinite amount of time. Since the student can swim 1.20 km in 1000 s, we should expect that the trip will definitely take longer than in still water, maybe about 2000 s (~30 minutes).

The total time in the river is the longer time upstream (against the current) plus the shorter time downstream (with the current). For each part, we will use the basic equation $t = d/v$, where v is the speed of the student relative to the shore.

$$t_{up} = \frac{d}{v_{student} - v_{stream}} = \frac{1000 \text{ m}}{1.20 \text{ m}/\text{s} - 0.500 \text{ m}/\text{s}} = 1430 \text{ s}$$

$$t_{dn} = \frac{d}{v_{student} + v_{stream}} = \frac{1000 \text{ m}}{1.20 \text{ m}/\text{s} + 0.500 \text{ m}/\text{s}} = 588 \text{ s}$$

Total time in river, $\quad t_{river} = t_{up} + t_{dn} = 2.02 \times 10^3 \text{ s}$ ◊

In still water, $\quad t_{still} = \frac{d}{v} = \frac{2000 \text{ m}}{1.20 \text{ m}/\text{s}} = 1.67 \times 10^3 \text{ s} \quad \text{or} \quad t_{still} = 0.827 t_{river}$ ◊

As we predicted, it does take the student longer to swim up and back in the moving stream than in still water (21% longer in this case), and the amount of time agrees with our estimation.

37. A science student is riding on a flatcar of a train traveling along a straight horizontal track at a constant speed of 10.0 m/s. The student throws a ball into the air along a path that he judges to make an initial angle of 60.0° with the horizontal and to be in line with the track. The student's professor, who is standing on the ground nearby, observes the ball to rise vertically. How high does she see the ball rise?

Solution Shown on the right, $\quad \mathbf{v}_{be} = \mathbf{v}_{ce} + \mathbf{v}_{bc}$

with $\quad \mathbf{v}_{bc}$ = the velocity of the ball relative to the car

$\mathbf{v}_{be}$ = the velocity of the ball relative to the Earth

and $\quad \mathbf{v}_{ce}$ = car velocity relative to the Earth = 10.0 m/s

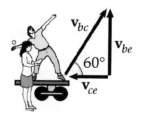

From the figure, we have

$$v_{ce} = v_{bc} \cos 60.0°$$

So

$$v_{bc} = \frac{10.0 \text{ m}/\text{s}}{\cos 60.0°} = 20.0 \text{ m}/\text{s}$$

Again from the figure,

$$v_{be} = v_{bc} \sin 60.0° + 0 = (20.0 \text{ m}/\text{s})(0.866) = 17.3 \text{ m}/\text{s}$$

This is the initial velocity of the ball relative to the Earth. Now we can calculate the maximum height that the ball rises:

From $v_{yf}^2 = v_{yi}^2 + 2ah,$

$$0 = (17.3 \text{ m}/\text{s})^2 + 2(-9.80 \text{ m}/\text{s}^2)h$$

$$h = 15.3 \text{ m} \qquad \Diamond$$

---

45. A home run is hit in such a way that the baseball just clears the top row of bleachers, 21.0 m high, located 130 m from home plate. The ball is hit at an angle of 35.0° to the horizontal, and air resistance is negligible. Find (a) the initial speed of the ball, (b) the time it takes the ball to reach the cheap seats, and (c) the velocity components and the speed of the ball when it passes over the top row. (Assume the ball is hit at a height of 1.00 m above the ground.)

Solution   Let the initial speed of the ball be v_i, and the initial angle $\theta_i = 35.0°$. Set the starting point at the origin:

$$(x_i, y_i) = (0 \text{ m}, 0 \text{ m})$$

When

$$x_f = 130 \text{ m}, \quad y_f = 20.0 \text{ m}$$

$$x_f = x_i + v_{xi}t = v_i t \cos 35.0°$$

$$v_i t \cos 35.0° = 130 \text{ m}$$

and

$$v_i t = \frac{130 \text{ m}}{\cos 35.0°} = 158.7 \text{ m}$$

$$v_{xi} = v_i \cos 35.0°$$
$$v_{yi} = v_i \sin 35.0°$$

(130 m, 20.0 m)

Next, $$y_f = v_{yi}t - \tfrac{1}{2}gt^2$$

$$y_f = (v_i \sin 35.0°)t - \left(\tfrac{1}{2}\right)(9.80 \text{ m/s}^2)t^2$$

Substituting for v_it, $$20.0 \text{ m} = (158.7 \text{ m})(\sin 35.0°) - \left(\tfrac{1}{2}\right)(9.80 \text{ m/s}^2)t^2$$

(b) and $$t = \sqrt{\frac{71.0 \text{ m}}{4.90 \text{ m/s}^2}} = 3.81 \text{ s}$$ ◊

(a) Therefore, $$v_i = \frac{158.7 \text{ m}}{3.81 \text{ s}} = 41.7 \text{ m/s}$$ ◊

(c) $$v_{yf} = v_{yi} - gt = v_i \sin 35.0° - gt = (23.9 \text{ m/s}) - \left(9.80 \text{ m/s}^2\right)t$$

At $t = 3.81$ s, $$v_y = -13.4 \text{ m/s}$$ ◊

$$v_{xf} = v_{xi} = v_i \cos 35.0° = (41.7 \text{ m/s}) \cos 35.0° = 34.1 \text{ m/s}$$ ◊

and $$|\mathbf{v}| = \sqrt{v_x^2 + v_y^2} = \sqrt{34.1^2 + (-13.4)^2} = 36.6 \text{ m/s}$$ ◊

---

53. A car is parked on a steep incline overlooking the ocean. The incline makes an angle of 37.0° below the horizontal. The negligent driver leaves the car in neutral, and the parking brakes are defective. The car rolls from rest down the incline with a constant acceleration of 4.00 m/s², traveling 50.0 m to the edge of a vertical cliff. The cliff is 30.0 m above the ocean. Find (a) the speed of the car when it reaches the edge of the cliff and the time it takes to get there, (b) the velocity of the car when it lands in the ocean, (c) the total time the car is in motion, and (d) the position of the car when it lands in the ocean, relative to the base of the cliff.

Solution   From point A to point B (along the incline), the car can be modeled as a particle under constant acceleration in one dimension, starting from rest ($v_i = 0$). Therefore, taking s to be the position along the incline,

(a) $v_B{}^2 = v_i{}^2 + 2a(s_f - s_i) = 2a(s_B - s_A)$

$v_B = \sqrt{(2)(4.00 \text{ m/s}^2)(50.0 \text{ m})} = 20.0 \text{ m/s}$ ◊

We can find the elapsed time from $v_B = v_i + at$:

$t_{AB} = \dfrac{v_B - v_i}{a} = \dfrac{20.0 \text{ m/s} - 0}{4.00 \text{ m/s}^2} = 5.00 \text{ s}$ ◊

(b)   After the car passes the top of the cliff, it becomes a projectile. At the edge of the cliff, the components of velocity v_B are:

$v_{yB} = (-20.0 \text{ m/s}) \sin 37.0° = -12.0 \text{ m/s}$ $v_{xB} = (20.0 \text{ m/s}) \cos 37.0° = 16.0 \text{ m/s}$

There is no further horizontal acceleration, so $v_{xC} = v_{xB} = 16.0 \text{ m/s}$

However, the downward (negative) vertical velocity is affected by free fall:

$v_{yC} = \pm\sqrt{2a_y(\Delta y) + v_{yB}{}^2} = \pm\sqrt{2(-9.80 \text{ m/s}^2)(-30.0 \text{ m}) + (-12.0 \text{ m/s})^2} = -27.1 \text{ m/s}$

$\mathbf{v}_C = 16.0\mathbf{i} - 27.1\mathbf{j} \text{ m/s}$ ◊

(c)   From point B to C, the time $t_{BC} = \dfrac{v_{yC} - v_{yB}}{a_y} = \dfrac{(-27.1 \text{ m/s}) - (-12.0 \text{ m/s})}{(-9.80 \text{ m/s}^2)} = 1.54 \text{ s}$

The total elapsed time is $t_{AC} = t_{AB} + t_{BC} = 6.54 \text{ s}$ ◊

(d)   The horizontal distance covered is $\Delta x = v_{xB}t_{BC} = (16.0 \text{ m/s})(1.54 \text{ s}) = 24.6 \text{ m}$ ◊

55. A skier leaves the ramp of a ski jump with a velocity of 10.0 m/s, 15.0° above the horizontal, as in Figure P3.55. The slope is inclined at 50.0°, and air resistance is negligible. Find (a) the distance from the ramp to where the jumper lands and (b) the velocity components just before the landing. (How do you think the results might be affected if air resistance were included? Note that jumpers lean forward in the shape of an airfoil, with their hands at their sides, to increase their distance. Why does this work?)

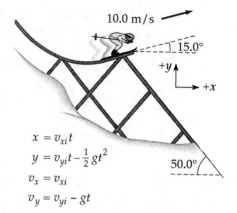

$$x = v_{xi}t$$
$$y = v_{yi}t - \tfrac{1}{2}gt^2$$
$$v_x = v_{xi}$$
$$v_y = v_{yi} - gt$$

Solution   Set point '0' where the skier takes off, and point '2' where the skier lands. Define the coordinate system as shown in the figure.

Figure P3.55

$$v_i = 10.0 \text{ m/s at } 15.0°$$
$$v_{xi} = v_i \cos \theta_i = (10.0 \text{ m/s}) \cos 15.0° = 9.66 \text{ m/s}$$
$$v_{yi} = v_i \sin \theta_i = (10.0 \text{ m/s}) \sin 15.0° = 2.59 \text{ m/s}$$

(a) (1) The skier travels horizontally $x_2 - x_0 = v_{xi}t = (9.66 \text{ m/s})t$

(2) and vertically $y_2 - y_0 = v_{yi}t - \tfrac{1}{2}gt^2 = (2.59 \text{ m/s})t - (4.90 \text{ m/s}^2)t^2$

(3) The skier hits the slope when $\dfrac{y_2 - y_0}{x_2 - x_0} = \tan(-50.0°) = -1.19$

Substituting (1) and (2) into (3), $\dfrac{(2.59 \text{ m / s})t - (4.90 \text{ m / s}^2)t^2}{(9.66 \text{ m / s})t} = -1.19$

Since we ignore the solution $t = 0$, $-4.90t + 14.1 = 0$

and $t = 2.88$ s. Solving (1), $x_2 - x_0 = (9.66 \text{ m/s})t = 27.8 \text{ m}$

By Figure P3.55 $d = \dfrac{x_2 - x_0}{\cos 50.0°} = \dfrac{27.8 \text{ m}}{\cos 50.0°} = 43.2 \text{ m}$ ◊

(b) The horizontal final velocity is simply $v_{xf} = v_{xi} = 9.66 \text{ m/s}$ ◊

The vertical component is found from $v_{yf} = v_{yi} - gt = 2.59 \text{ m/s} - (9.80 \text{ m/s}^2)t$

When $t = 2.88$ s, $v_{yf} = -25.6 \text{ m/s}$ ◊

The 'drag' force of air resistance would necessarily decrease both components of the ski jumper's impact velocity. On the other hand, the lift force of the air could extend her time of flight and increase the distance of her jump. If the jumper has the profile of an airplane wing, she can deflect downward the air through which she passes, to make the air deflect her upward.

Chapter 4

The Laws of Motion

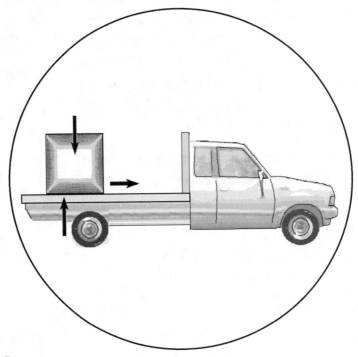

INTRODUCTION

In Chapters 2 and 3, we described the motion of particles based on the definitions of displacement, velocity, and acceleration. In this chapter, we use the concepts of force and mass to describe the change in motion of particles. We then discuss the three basic laws of motion, which are based on experimental observations and were formulated more than three centuries ago by Newton.

Classical mechanics describes the relationship between the motion of a body and the forces acting on that body. Classical mechanics deals only with objects that (a) are large compared with the dimensions of atoms and (b) move at speeds that are much less than the speed of light.

We learn in this chapter how it is possible to describe an object's acceleration in terms of the resultant force acting on the object and its mass. This force represents the interaction of the object with its environment. Mass is a measure of the object's inertia, that is, its tendency to resist an acceleration when a force acts on it.

We also discuss force laws, which describe the quantitative method of calculating the force on an object if its environment is known. These force laws, together with the laws of motion, are the foundations of classical mechanics.

NOTES FROM SELECTED CHAPTER SECTIONS

4.1   The Concept of Force

Equilibrium is the condition under which the **net force** (vector sum of all forces) acting on an object is zero. An object in equilibrium has a zero acceleration (velocity is constant or equals zero).

Fundamental forces in nature are: (1) gravitational (attractive forces between objects due to their masses), (2) electromagnetic forces (between electric charges at rest or in motion), (3) nuclear forces (between subatomic particles), and (4) weak forces (accompanying the process of radioactive decay).

Classical physics is concerned with contact forces (which are the result of physical contact between two or more objects) and field forces (which act through empty space and do not involve physical contact).

4.2   Newton's First Law

Newton's first law is called the **law of inertia** and states that an object at rest will remain at rest and an object in motion will remain in motion with a constant velocity unless acted on by a **net external force.**

Mass and **weight** are two different physical quantities. The weight of a body is equal to the **force of gravity** acting on the body and varies with location in the Earth's gravitational field. Mass is an **inherent property** of a body and is a measure of the body's inertia (resistance to change in its state of motion). The SI unit of mass is the **kilogram** and the unit of weight is the **newton.**

4.3   Inertial Mass

Inertial mass is a measure of an object's resistance to a change in motion, in response to an external force. Mass is an inherent property of an object, independent of the surroundings. Mass and weight are different quantities; the weight of an object is equal to the magnitude of the gravitational force exerted on the object.

4.4   Newton's Second Law — The Particle Under a Net Force

Newton's second law, the law of **acceleration**, states that the acceleration of an object is directly proportional to the net force acting on it and inversely proportional to its mass. The direction of the acceleration is the direction of the net force. The net force is also called the resultant force, total force, or sum of all forces.

4.6   Newton's Third Law

Newton's third law, the law of **action-reaction**, states that when two bodies interact, the force which body "1" exerts on body "2" (the **action force**) is equal in magnitude and opposite in direction to the force which body "2" exerts on body "1" (the **reaction force**). A consequence of the third law is that forces occur in **pairs.** Remember that the action force and the reaction force act on **different objects.**

4.7   Applications of Newton's Laws

Construction of a **free-body diagram** is an important step in the application of Newton's laws of motion to the solution of problems involving bodies in equilibrium or accelerating under the action of external forces. The diagram should include an arrow labeled to identify each of the external forces acting on the body whose motion (or condition of equilibrium) is to be studied. Forces which are the **reactions** to these external forces must **not** be included. When a system consists of more than one body or mass, you must construct a free-body diagram for each mass.

EQUATIONS AND CONCEPTS

A quantitative measurement of mass (the term used to measure inertia) can be made by comparing the accelerations that a given force will produce on different bodies. If a given force acting on a body of mass m_1 produces an acceleration a_1 and the same force acting on a body of mass m_2 produces an acceleration a_2, the ratio of the two masses equals the inverse of the ratio of the two accelerations.

$$\frac{m_1}{m_2} = \frac{a_2}{a_1} \qquad (4.1)$$

The acceleration of an object is proportional to the net force acting on it and inversely proportional to its mass. This is a statement of Newton's second law.

$$\Sigma\mathbf{F} = m\mathbf{a} \qquad (4.2)$$

When several forces act on an object, it is often convenient to write the equation expressing Newton's second law as component equations. The orientation of the coordinate system can often be chosen so that the object has a nonzero acceleration along only one direction.

$$\Sigma F_x = m a_x \qquad (4.3)$$
$$\Sigma F_y = m a_y$$
$$\Sigma F_z = m a_z$$

Calculations with Equation 4.3 must be made using a consistent set of units for the quantities' force, mass, and acceleration. The SI unit of force is the newton (N), defined as the force that, when acting on a 1-kg mass, produces an acceleration of 1 m/s^2.

$$1\,\mathrm{N} \equiv 1\,\mathrm{kg \cdot m/s^2} \qquad (4.4)$$

Weight is not an inherent property of a body, but depends on the local value of g and varies with location.

$$F_g = mg \qquad (4.5)$$

Newton's third law states that the force exerted on body 1 by body 2 is equal in magnitude and opposite in direction to the force exerted on body 2 by body 1.

$$\mathbf{F}_{12} = -\mathbf{F}_{21} \qquad (4.6)$$

SUGGESTIONS, SKILLS, AND STRATEGIES

The following procedure is recommended when dealing with problems involving the application of Newton's second law:

- Draw a simple, neat diagram of the situation.

- Isolate the object of interest whose motion is being analyzed. Draw a free-body diagram for this object; that is, a diagram showing **all external forces acting on the object.** For systems containing more than one object, draw **separate** diagrams for each object. Do not include forces that the object exerts on its surroundings.

- Establish convenient coordinate axes for each object and find the components of the forces along these axes.

- Apply Newton's second law, $\Sigma \mathbf{F} = m\mathbf{a}$, in the x and y directions for each object under consideration. For cases when the object is in equilibrium along either direction, $\Sigma F = 0$ along that direction.

- Solve the component equations for the unknowns. Remember that you must have as many independent equations as you have unknowns in order to obtain a complete solution.

- Often in solving such problems, one must also use the equations of kinematics (motion with constant acceleration) to find all the unknowns.

REVIEW CHECKLIST

▷ State in your own words a description of Newton's laws of motion, recall physical examples of each law, and identify the action-reaction force pairs in a multiple-body interaction problem as specified by Newton's third law.

▷ Apply Newton's laws of motion to various mechanical systems using the recommended procedure discussed in Section 4.7. Most important, you should identify all external forces acting on the system, draw the **correct** free-body diagrams which apply to each body of the system, and apply Newton's second law, $\mathbf{F} = m\mathbf{a}$, in **component** form.

▷ Apply the equations of kinematics (which involve the quantities position, velocity, and acceleration) as described in Chapter 2 along with those methods and equations of Chapter 4 (involving mass, force, and acceleration) to the solutions of problems where **both** the kinematic and dynamic aspects are present.

▷ Be familiar with solving several linear equations simultaneously for the unknown quantities. Recall that you must have as many **independent** equations as you have unknowns.

ANSWERS TO SELECTED CONCEPTUAL QUESTIONS

2. In the motion picture *It Happened One Night* (Columbia Pictures, 1934), Clark Gable is standing inside a stationary bus in front of Claudette Colbert, who is seated. The bus suddenly starts moving forward and Clark falls into Claudette's lap. Why did this happen?

Answer When the bus starts moving, the mass of Claudette is accelerated by the force of the back of the seat on her body. Clark is standing, however, and the only force on him is the friction between his shoes and the floor of the bus. Thus, when the bus starts moving, his feet start accelerating forward, but the rest of his body experiences almost no accelerating force (only that due to his being attached to his accelerating feet!). As a consequence, his body tends to stay almost at rest, according to Newton's first law, relative to the ground. Relative to Claudette, however, he is moving toward her and falls into her lap. (Both performers won Academy Awards.)

7. Identify the action - reaction pairs in the following situations: a man takes a step; a snowball hits a girl in the back; a baseball player catches a ball; a gust of wind strikes a window.

Answer As a man takes a step, the action is the force his foot exerts on the Earth; the reaction is the force of the Earth on his foot. In the second case, the action is the force exerted on the girl's back by the snowball; the reaction is the force exerted on the snowball by the girl's back. The third action is the force of the glove on the ball; the reaction is the force of the ball on the glove. The fourth action is the force exerted on the window by the air molecules; the reaction is the force on the air molecules exerted by the window.

□   □   □   □

12. A weight lifter stands on a bathroom scale. He pumps a barbell up and down. What happens to the reading on the bathroom scale as this is done? Suppose he is strong enough to actually **throw** the barbell upward. How does the reading on the scale vary now?

Answer If the barbell is not moving, the reading on the bathroom scale is the combined weight of the weightlifter and the barbell. At the beginning of the lift of the barbell, the barbell accelerates upward. By Newton's third law, the barbell pushes downward on the hands of the weightlifter with more force than its weight, in order to accelerate. As a result, he is pushed with more force into the scale, increasing its reading. Near the top of the lift, the weightlifter reduces the upward force, so that the acceleration of the barbell is downward, causing it to come to rest. While the barbell is coming to rest, it pushes with less force on the weightlifter's hands, so the reading on the scale is below the combined stationary weight. If the barbell is held at rest for a moment at the top of the lift, the scale reading is simply the combined weight. As it begins to be brought down, the reading decreases, as the force of the weightlifter on the barbell is reduced. The reading increases as the barbell is slowed down at the bottom.

If we now consider the throwing of the barbell, we have the same behavior as before, except that the variations in scale reading will be larger, since more force must be applied to throw the barbell upward rather than just lift it. Once the barbell leaves the weightlifter's hands, the reading will suddenly drop to just the weight of the weightlifter, and will rise suddenly when the barbell is caught.

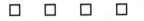

SOLUTIONS TO SELECTED END-OF-CHAPTER PROBLEMS

7. Two forces, F_1 and F_2, act on a 5.00-kg object. If $F_1 = 20.0$ N and $F_2 = 15.0$ N, find the accelerations in (a) and (b) of Figure P4.7.

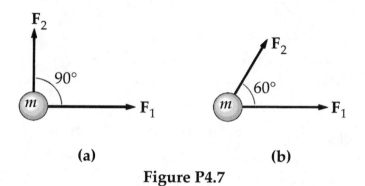

(a)   (b)

Figure P4.7

Solution   We use the particle under a net force model:

$m = 5.00$ kg

(a) $\Sigma \mathbf{F} = \mathbf{F}_1 + \mathbf{F}_2 = (20.0\mathbf{i} + 15.0\mathbf{j})$ N

$$\mathbf{a} = \frac{\Sigma \mathbf{F}}{m} = (4.00\mathbf{i} + 3.00\mathbf{j}) \text{ m/s}^2 = 5.00 \text{ m/s}^2 \text{ at } 36.9° \qquad \Diamond$$

(b) $\Sigma \mathbf{F} = \mathbf{F}_1 + \mathbf{F}_2 = [20.0\mathbf{i} + (15.0 \cos 60° \, \mathbf{i} + 15.0 \sin 60° \, \mathbf{j})]$ N $= (27.5\mathbf{i} + 13.0\mathbf{j})$ N

$$\mathbf{a} = \frac{\Sigma \mathbf{F}}{m} = (5.50\mathbf{i} + 2.60\mathbf{j}) \text{ m/s}^2 = 6.08 \text{ m/s}^2 \text{ at } 25.3° \qquad \Diamond$$

13. An electron of mass 9.11×10^{-31} kg has an initial speed of 3.00×10^5 m/s. It travels in a straight line, and its speed increases to 7.00×10^5 m/s in a distance of 5.00 cm. Assuming its acceleration is constant, (a) determine the force on the electron and (b) compare this force with the weight of the electron, which we ignored.

Solution   We know the initial and final velocities, and the distance involved. From $v_f^2 = v_i^2 + 2ax$ and $\Sigma F = ma$ we can solve for the acceleration and the force.

$$a = \frac{\left(v_f^2 - v_i^2\right)}{2x} \quad \text{and} \quad \Sigma F = \frac{m\left(v_f^2 - v_i^2\right)}{2x}$$

(a) $\Sigma F = \dfrac{\left(9.11\times10^{-31}\ \text{kg}\right)\left(\left(7.00\times10^5\ \text{m/s}\right)^2 - \left(3.00\times10^5\ \text{m/s}\right)^2\right)}{(2)(0.0500\ \text{m})} = 3.64\times10^{-18}\ \text{N}$ ◊

(b)   The weight of the electron is

$$F = mg = \left(9.11\times10^{-31}\ \text{kg}\right)\left(9.80\ \text{m / s}^2\right) = 8.93\times10^{-30}\ \text{N}$$

The ratio of the accelerating force to the weight is 4.08×10^{11}.   ◊

---

21.   A bag of cement of weight F_g hangs from three wires as shown in Figure P4.20. Two of the wires make angles θ_1 and θ_2 with the horizontal. If the system is in equilibrium, show that the tension in the left-hand wire is

$$T_1 = \frac{F_g \cos \theta_2}{\sin (\theta_1 + \theta_2)}$$

Solution

We use the particle in equilibrium model. Draw a free-body diagram for the knot where the three ropes are joined. Choose the x axis to be horizontal and apply Newton's second law in component form.

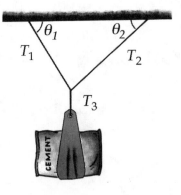

Figure P4.20

$\Sigma F_x = 0;$ $\quad T_2 \cos\theta_2 - T_1 \cos\theta_1 = 0$ (1)

$\Sigma F_y = 0;$ $\quad T_2 \sin\theta_2 + T_1 \sin\theta_1 - F_g = 0$ (2)

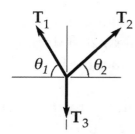

Solve equation (1) for $\qquad T_2 = \dfrac{T_1 \cos\theta_1}{\cos\theta_2}$

Substitute this expression for T_2 into Equation (2):

$$\left(\frac{T_1 \cos\theta_1}{\cos\theta_2}\right) \sin\theta_2 + T_1 \sin\theta_1 = F_g$$

Solve for $\qquad T_1 = \dfrac{F_g \cos\theta_2}{\cos\theta_1 \sin\theta_2 + \sin\theta_1 \cos\theta_2}$

Use trigonometric identity, $\qquad \sin(\theta_1 + \theta_2) = \cos\theta_1 \sin\theta_2 + \sin\theta_1 \cos\theta_2$

to find $\qquad T_1 = \dfrac{F_g \cos\theta_2}{\sin(\theta_1 + \theta_2)}$ ◊

---

23. A 1.00-kg object is observed to accelerate at 10.0 m/s² in a direction 30.0° north of east (Fig. P4.23). The force **F**$_2$ acting on the mass has a magnitude of 5.00 N and is directed north. Determine the magnitude and direction of the force **F**$_1$ acting on the mass.

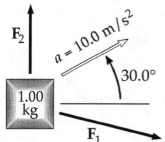

Figure P4.23

Solution

Choose directions east and north along **i** and **j**, respectively. The acceleration, broken into its components, is:

$$\mathbf{a} = [(10.0 \cos 30.0°)\mathbf{i} + (10.0 \sin 30.0°)\mathbf{j}] \, \text{m}/\text{s}^2 = (8.66\mathbf{i} + 5.00\mathbf{j}) \, \text{m}/\text{s}^2$$

From Newton's law, $\qquad\qquad\qquad\qquad \mathbf{F}_{net} = m\mathbf{a} = (8.66\mathbf{i} + 5.00\mathbf{j}) \, \text{N}$

But since the net force is equal to the sum of the forces, $\qquad \mathbf{F}_{net} = \mathbf{F}_1 + \mathbf{F}_2$

and $\qquad \mathbf{F}_1 = \mathbf{F}_{net} - \mathbf{F}_2 = (8.66\mathbf{i} + 5.00\mathbf{j} - 5.00\mathbf{j}) \, \text{N} = 8.66\mathbf{i} \, \text{N} = 8.66 \, \text{N, due east}$ ◊

29. A block is given an initial speed of 5.00 m/s up a frictionless 20.0° incline (see Figure P4.28). How far up the incline does the block slide before coming to rest?

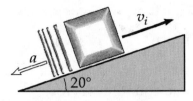

Figure P4.28

Solution Every successful physics student (this means you) learns to solve inclined-plane problems.

Hint one: Try to set the axes in the direction of motion. In this case, take the x-axis along the incline, so that $a_y = 0$.

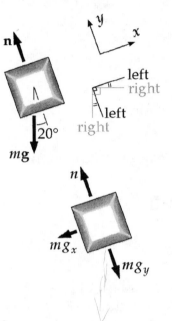

Hint two: Recognize that the 20.0° angle between the x-axis and horizontal implies a 20.0° angle between the weight vector and the y-axis. Why? Because "angles are equal if their sides are perpendicular, right side to right side and left side to left side." Either you learned this theorem in geometry class, or you learn it now, since it is a theorem used often in physics.

Hint three: The 20.0° angle lies between $m\mathbf{g}$ and the y-axis, so split the weight vector into its x and y components:

$$mg_x = -mg\sin 20.0° \qquad mg_y = -mg\cos 20.0°$$

Now, Newton's law applies for **each** axis. Applying it to the x axis,

$$\Sigma F_x = ma_x: \qquad -mg\ \sin 20° = ma_x$$

$$a_x = -g\sin 20° = -\left(9.80\ \text{m/s}^2\right)\sin 20° = -3.35\ \text{m/s}^2$$

From Eq. 2.12, $\qquad v_{xf}^{\ 2} = v_{xi}^{\ 2} + 2a_x\left(x_f - x_i\right)$

$$0 = \left(5.00\ \text{m/s}\right)^2 + 2\left(-3.35\ \text{m/s}^2\right)\left(x_f - x_i\right)$$

Solving, $\qquad \left(x_f - x_i\right) = 3.73\ \text{m}$ ◊

33. In the system shown in Figure P4.33, a horizontal force of magnitude F_x acts on the 8.00-kg mass. The horizontal surface is frictionless. (a) For what values of F_x does the 2.00-kg mass accelerate upward? (b) For what values of F_x is the tension in the cord zero? (c) Plot the acceleration of the 8.00-kg object versus F_x. Include values of F_x from –100 N to +100 N.

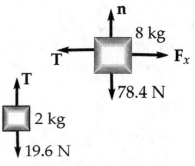

Figure P4.33

Solution The blocks' weights are:

$$F_{g1} = m_1g = (8.00 \text{ kg})(9.80 \text{ m}/\text{s}^2) = 78.4 \text{ N}$$
$$F_{g2} = m_2g = (2.00 \text{ kg})(9.80 \text{ m}/\text{s}^2) = 19.6 \text{ N}$$

Let T be the tension in the connecting cord and draw a free-body diagram for each block.

(a) For the 2-kg mass, with the y-axis directed upwards,

$$\Sigma F_y = m a_y \qquad \text{yields} \qquad T - 19.6 \text{ N} = (2.00 \text{ kg})a_y \qquad (1)$$

Thus, we find that $a_y > 0$ when $T > 19.6 \text{ N}$. For acceleration by the system of two blocks, $F_x \geq T$, so $F_x > 19.6 \text{ N}$ whenever the 2-kg mass accelerates upward. ◊

(b) Looking at the free-body diagram for the 8.00-kg mass, and taking the $+x$ direction to be directed to the right, we can apply Newton's law in the horizontal direction:

From $\Sigma F_x = m a_x$ $\qquad\qquad -T + F_x = (8.00 \text{ kg})a_x \qquad (2)$

If $T = 0$, and $a_x \leq -9.80 \text{ m}/\text{s}^2$ $\qquad\qquad F_x \leq -78.4 \text{ N}$ ◊

(c) If $F_x \geq -78.4 \text{ N}$, then both equations (1) and (2) apply. Substituting the value for T from the (1) into (2),

$$-(2.00 \text{ kg})a_y - 19.6 \text{ N} + F_x = (8.00 \text{ kg})a_x$$

In this case $a_x = a_y$: $\qquad\qquad F_x = (8.00 \text{ kg} + 2.00 \text{ kg})a_x + 19.6 \text{ N}$

$$a_x = \frac{F_x}{10.0 \text{ kg}} - 1.96 \text{ m} / \text{s}^2 \qquad\qquad (F_x \geq -78.4 \text{ N}) \qquad\qquad (3)$$

From part (b), we find that if $F_x \leq -78.4$ N, then $T = 0$ and equation (2) becomes $F_x = (8.00 \text{ kg})a_x$:

$$a_x = F_x/8.00 \text{ kg} \qquad\qquad (F_x \leq -78.4 \text{ N}) \qquad\qquad (4)$$

Observe that we have translated the pictorial representation into a simplified pictorial representation and then into a mathematical representation. We proceed to a tabular representation and a graphical representation of equations (3) and (4):

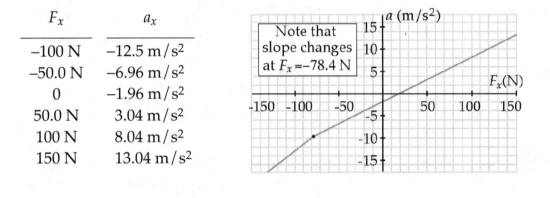

F_x	a_x
−100 N	−12.5 m/s²
−50.0 N	−6.96 m/s²
0	−1.96 m/s²
50.0 N	3.04 m/s²
100 N	8.04 m/s²
150 N	13.04 m/s²

35. A 72.0-kg man stands on a spring scale in an elevator. Starting from rest, the elevator ascends, attaining its maximum speed of 1.20 m/s in 0.800 s. It travels with this constant speed for the next 5.00 s. The elevator then undergoes a uniform acceleration in the negative y direction for 1.50 s and comes to rest. What does the spring scale register (a) before the elevator starts to move? (b) during the first 0.800 s? (c) while the elevator is traveling at constant speed? (d) during the time it is slowing down?

Solution

We use the particle under constant acceleration and particle under net force models in parts (b) and (d), and use the particle under constant velocity and particle in equilibrium models in parts (a) and (c). Let S be the weight registered by the scale. In each case draw a free-body diagram showing forces acting on the man and apply Newton's second law.

(a) Before the elevator starts to move:

$a = 0$ and $\Sigma F_y = S - mg = 0$

$S = mg = (72.0 \text{ kg})(9.80 \text{ m/s}^2) = 706 \text{ N}$ ◊

(b) $a = \dfrac{v_f - v_i}{t} = \dfrac{1.20 \text{ m/s} - 0}{0.800 \text{ s}} = 1.50 \text{ m/s}^2$

During upward acceleration, the scale reading is $S = mg + ma$, because the man's feet are pressed against the floor by the elevator's upward acceleration, as well as by his weight. We also find this is true by applying the force equation, with S equal to the normal force:

$\Sigma F_y = S - mg = ma$

$S = m(a + g) = (72.0 \text{ kg})(1.50 \text{ m/s}^2 + 9.80 \text{ m/s}^2) = 814 \text{ N}$ ◊

(c) While traveling at constant speed, $a = 0$; and therefore the force registered by the scale will be the same as in part (a):

$S = mg = 706 \text{ N}$ ◊

(d) While slowing down: $a = \dfrac{v_f - v_i}{t} = \dfrac{0 - 1.20 \text{ m/s}}{1.50 \text{ s}} = -0.800 \text{ m/s}^2$

so $\Sigma F = S - mg = ma$

$S = m(a + g) = (72.0 \text{ kg})(-0.800 \text{ m/s}^2 + 9.80 \text{ m/s}^2) = 648 \text{ N}$ ◊

37. To model a spacecraft, a toy rocket engine is securely fastened to a large puck, which can glide with negligible friction over a horizontal surface, taken as the xy plane. The 4.00-kg puck has a velocity of 3.00i m/s at one instant. Eight seconds later, its velocity is to be $(8.00\mathbf{i} + 10.0\mathbf{j})$ m/s. Assuming the rocket engine exerts a constant horizontal force, find (a) the components of the force and (b) its magnitude.

Solution

We use the particle under constant acceleration and particle under net force models. We first calculate the acceleration of the puck:

$$\bar{\mathbf{a}} = \frac{\Delta\mathbf{v}}{\Delta t} = \frac{(8.00\mathbf{i} + 10.0\mathbf{j}) \text{ m/s} - 3.00\mathbf{i} \text{ m/s}}{8.00 \text{ s}} = 0.625\mathbf{i} \text{ m/s}^2 + 1.25\mathbf{j} \text{ m/s}^2$$

In $\Sigma\mathbf{F} = m\mathbf{a}$, the only horizontal force is the thrust $\mathbf{F}$ of the rocket:

$$\mathbf{F} = (4.00 \text{ kg})(0.625\mathbf{i} \text{ m/s}^2 + 1.25\mathbf{j} \text{ m/s}^2) = 2.50\mathbf{i} \text{ N} + 5.00\mathbf{j} \text{ N} \qquad \Diamond$$

Then, $\quad |\mathbf{F}| = \sqrt{(2.50 \text{ N})^2 + (5.00 \text{ N})^2} = 5.59 \text{ N} \qquad \Diamond$

---

39. An inventive child named Pat wants to reach an apple in a tree without climbing the tree. Sitting in a chair connected to a rope that passes over a frictionless pulley (Fig. P4.39), Pat pulls on the loose end of the rope with such a force that the spring scale reads 250 N. Pat's weight is 320 N, and the chair weighs 160 N. (a) Draw free-body diagrams for Pat and the chair considered as separate systems and another diagram for Pat and the chair considered as one system. (b) Show that the acceleration of the system is upward and find its magnitude. (c) Find the force Pat exerts on the chair.

Figure P4.39

Solution

(a)

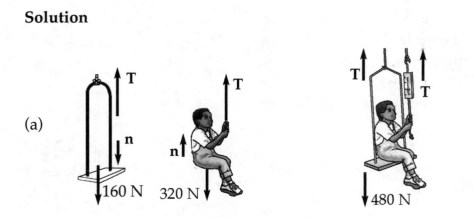

$\Diamond$

(b) First consider Pat and the chair as the system. Note that **two** ropes support the system, and $T = 250$ N in each rope. Applying $\Sigma F = ma$,

$2T - (160 + 320)$ N $= ma$ where $m = \dfrac{480 \text{ N}}{9.80 \text{ m}/\text{s}^2} = 49.0$ kg

Solving for a gives $a = \dfrac{(500 - 480) \text{ N}}{49.0 \text{ kg}} = 0.408 \text{ m}/\text{s}^2$ $\Diamond$

(c) On Pat, we apply $\Sigma F = ma$: $n + T - 320$ N $= ma$

where $m = \dfrac{320 \text{ N}}{9.80 \text{ m}/\text{s}^2} = 32.7$ kg

$n = ma + 320 \text{ N} - T = 32.7 \text{ kg} (0.408 \text{ m}/\text{s}^2) + 320 \text{ N} - 250 \text{ N} = 83.3 \text{ N}$ $\Diamond$

47. An object of mass M is held in place by an applied force **F** and a pulley system as shown in Figure P4.47. The pulleys are massless and frictionless. Find (a) the tension in each section of rope, T_1, T_2, T_3, T_4, and T_5, and (b) the magnitude of **F**. (**Hint:** draw a free-body diagram for each pulley)

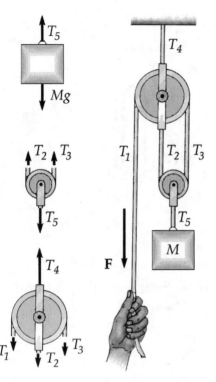

Solution   Draw free-body diagrams and apply Newton's 2nd law. (All forces are along the y axis.)

For M, $\Sigma F = 0 = T_5 - Mg$:

$$T_5 = Mg \qquad \Diamond$$

Assume frictionless pulleys. The tension is constant throughout a light, continuous rope.

Therefore, $T_1 = T_2 = T_3$

For the bottom pulley, $\Sigma F = 0 = T_2 + T_3 - T_5$

So $2T_2 = T_5$, and $T_1 = T_2 = T_3 = \dfrac{Mg}{2} \qquad \Diamond$

$$F_A = T_1 = \frac{Mg}{2} \qquad \Diamond$$

For the top pulley, $\sum F = 0 = T_4 - T_1 - T_2 - T_3$

Solving, $T_4 = T_1 + T_2 + T_3 = \dfrac{3Mg}{2} \qquad \Diamond$

Figure P4.47

51. What horizontal force must be applied to the cart shown in Figure P4.51 in order that the blocks remain stationary relative to the cart? Assume all surfaces, wheels, and pulley are frictionless. (**Hint:** Note that the force exerted by the string accelerates m_1.)

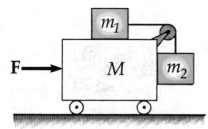

Figure P4.51

Solution

Draw separate free-body diagrams for blocks m_1 and m_2.

Remembering that normal forces are always perpendicular to the contacting surface, and always **push** on a body, draw $\mathbf{n}_1$ and $\mathbf{n}_2$ as shown. Note that m_2 should be in **contact** with the cart, and therefore does have a normal force from the cart.

Remembering that ropes always **pull** on bodies in the direction of the rope, draw the force exerted by the rope on each block. Its magnitude, the rope tension T, will be uniform if we model the rope as having negligible mass.

Finally, draw the gravitational force on each block, which always points downwards.

Use $\Sigma F = ma$ and the free-body diagrams above.

For m_2, $T - m_2 g = 0$ or $T = m_2 g$

For m_1, $T = m_1 a$ or $a = \dfrac{T}{m_1}$

Substituting for T, we have $a = \dfrac{m_2 g}{m_1}$

For all 3 blocks, $F = (M + m_1 + m_2)a.$

Therefore, $F = \left(M + m_1 + m_2\right)\left(\dfrac{m_2 g}{m_1}\right)$ ◊

53. A van accelerates down a hill (Fig. P4.53), going from rest to 30.0 m/s in 6.00 s. During the acceleration, a toy ($m = 0.100$ kg) hangs by a string from the van's ceiling. The acceleration is such that the string remains perpendicular to the ceiling. Determine (a) the angle θ and (b) the tension in the string.

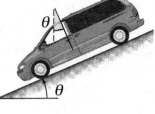

Figure P4.53

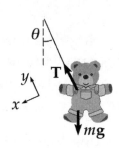

Solution   The acceleration is obtained from $v_f = v_i + at$:

$$30.0 \text{ m/s} = 0 + a(6.00 \text{ s})$$

$$a = 5.00 \text{ m/s}^2$$

The toy moves with the same acceleration as the van, 5.00 m/s^2 parallel to the hill. We take the x axis in this direction, so

$$a_x = 5.00 \text{ m/s}^2 \qquad \text{and} \qquad a_y = 0$$

The only forces on the toy are the string tension in the y direction and its weight, as shown in the free-body diagram.

$$mg = (0.100 \text{ kg})(9.80 \text{ m/s}^2) = 0.980 \text{ N}$$

Using $\Sigma F_x = ma_x$: $(0.980 \text{ N}) \sin \theta = (0.100 \text{ kg})(5.00 \text{ m/s}^2)$

(a) $\sin \theta = \dfrac{0.500}{0.980}$ and $\theta = 30.7°$ ◊

Using $\Sigma F_y = ma_y$: $+T - (0.980 \text{ N}) \cos \theta = 0$

(b) $T = (0.980 \text{ N}) \cos 30.7° = 0.843 \text{ N}$ ◊

Chapter 5

More Applications of Newton's Laws

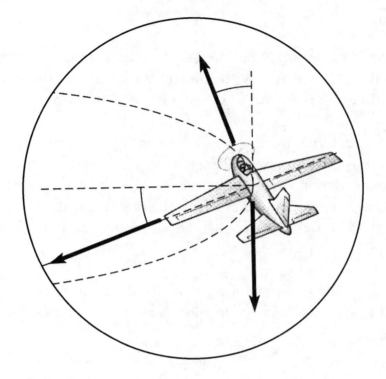

INTRODUCTION

In the previous chapter we introduced Newton's laws of motion and applied them to situations involving linear motion. In this chapter we shall apply Newton's laws of motion to circular motion. We shall also discuss sliding friction and the motion of an object through a viscous medium.

NOTES FROM SELECTED CHAPTER SECTIONS

5.1 Forces of Friction

When an object is in motion either on a surface or through a viscous medium such as air or water, the object reacts with the surface or the medium through which it is moving. The resulting resistance is called a **force of friction**.

In addition, if an external force is applied to an object at rest on a rough surface such that the force has a component directed parallel to the surface, there will be an opposing force of friction which is characteristic of the pair of surfaces. In this case, in which there is no relative motion between the object and the surface on which it rests, the force is called **static friction**. If relative motion occurs, the force is then one of **kinetic friction**.

Experiments show that the maximum value of the force of static friction ($f_{s,\,max}$) and the force of kinetic friction (f_k) are **proportional to the normal force between the two surfaces**. That is,

$$f_{s,\,max} = \mu_s n \qquad \text{and} \qquad f_k = \mu_k n$$

It should be noted that in general, for any pair of surfaces, μ_k is generally less than μ_s. In addition, the coefficients of friction are nearly independent of the area of contact between the surfaces.

5.2 Newton's Second Law Applied to a Particle in Uniform Circular Motion

If a particle moves in a circle of radius r with **constant speed**, it undergoes a centripetal acceleration v^2/r directed toward the center of rotation. (Recall that in this case, the centripetal acceleration arises from the change in **direction** of **v**.) Newton's second law applied to the motion says that the centripetal acceleration arises from some external, centripetal force acting toward the center of rotation.

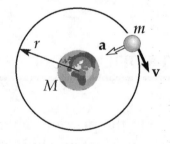

Figure 5.1

That is:

$$\sum \mathbf{F} \text{ (along } \hat{\mathbf{r}}) = m\mathbf{a}_r = -\frac{mv^2}{r}\hat{\mathbf{r}}$$

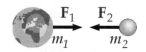

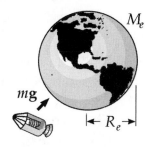

Figure 5.2

where $\hat{\mathbf{r}}$ is a unit vector pointing **radially outward** from the center.

The **universal gravitational constant**, G, is not to be confused with the acceleration g due to gravity. The gravitational force is always a force of attraction and, as shown in Figure 5.2, the force on m_1 due to m_2 is equal and opposite the force on m_2 due to m_1 (Newton's third law).

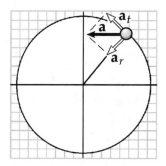

Figure 5.3

The gravitational force exerted by a spherically symmetric mass distribution on a particle outside the sphere is the **same** as if the entire mass of the sphere were concentrated at its center. The gravitational force of attraction on a mass m near the surface of the Earth is shown in Figure 5.3.

5.3   Nonuniform Circular Motion

If a particle moves in a circular path such that its **speed changes in time,** the particle also has a **tangential** component of acceleration $\mathbf{a}_t$, whose magnitude is dv/dt. In this case, the **total** acceleration $\mathbf{a}$ is the vector sum of $\mathbf{a}_r$ and $\mathbf{a}_t$.

$$\mathbf{a} = \mathbf{a}_r + \mathbf{a}_t$$

Figure 5.4

5.4   Motion in the Presence of Velocity-Dependent Resistive Forces

A body moving through a gas or liquid experiences a resistive force which can have a complicated velocity dependence. A falling body reaches a terminal velocity (maximum velocity) when the downward force of gravity is balanced by the upward resistive force. That is, when $\Sigma \mathbf{F} = 0$, $\mathbf{a} = 0$, and $\mathbf{v} = $ constant.

5.6   The Fundamental Forces of Nature

The **gravitational force** is the mutual force of attraction between any two objects in the universe; it is the weakest of the fundamental forces.

The **electromagnetic force** involves two types of particles: those with positive charge, and those with negative charge. It is this force that binds atoms and molecules to form matter.

The most important properties of electric charges can be characterized as follows:

- The two kinds of charges that exist in nature are labeled positive and negative. Charges of like sign repel each other, and charges of opposite sign attract each other.
- Possible values for charge are in discrete steps; in other words, charge is said to be quantized. Any isolated elementary charged particle has a charge of magnitude e. For example, electrons have a charge $-e$ and protons have a charge of $+e$. (Neutrons have no charge.) Since e is the fundamental unit of charge, the net charge of an object is Ne, where N is an integer. For macroscopic objects, N is very large and the quantization of charge is not noticed.
- Electric charge is always conserved. This means that in any kind of process--such as a collision event, a chemical reaction, or nuclear decay--the total charge of an isolated system remains constant.

The **nuclear force** is a very short range force which binds the nucleons to form a nucleus.

The **weak force** is much stronger than the gravitational force and much weaker than the strong force.

It is now known that the electromagnetic force and the weak force are both manifestations of a single force called the **electroweak force**.

EQUATIONS AND CONCEPTS

The force of static friction between two surfaces in contact but not in motion, relative to each other, cannot be greater than $\mu_s n$, where n is the normal (perpendicular) force between the two surfaces and μ_s (coefficient of static friction) is a dimensionless constant which depends on the nature of the pair of surfaces.

$$f_s \leq \mu_s n \qquad (5.1)$$

When two surfaces are in relative motion, the force of kinetic friction on each body is directed opposite to the direction of motion of that body relative to the other.

$$f_k = \mu_k n$$

When an object of mass m moves uniformly in a circular path, the net force acting on the object is a centripetal force (directed toward the center of the circular path.)

$$\Sigma F_r = m a_c = m \frac{v^2}{r} \qquad (5.3)$$

When Newton's second law is applied to the motion of an object falling vertically through a viscous medium, the motion can be described by a differential equation. If we consider a special case in which the resistive force is proportional to the velocity, $R = bv$, then the differential equation has a special form.

$$mg - bv = m \frac{dv}{dt}$$

or

$$\frac{dv}{dt} = g - \frac{b}{m} v$$

$\qquad (5.5)$

As the resistive force approaches the weight, the acceleration approaches zero, and the object reaches a terminal speed, v_t. Equation 5.6 gives the speed as a function of time, when the object is released from rest at $t = 0$.

$$v = v_T \left(1 - e^{-t/\tau} \right) \qquad (5.6)$$

$$\text{where} \quad v_T = \frac{mg}{b}$$

$$\text{and} \quad \tau = \frac{m}{b}$$

At high speed, resistance is due more to collision with the air than viscosity, and is proportional to v^2 instead of v.

$$R = \frac{1}{2} D \rho A v^2 \qquad (5.7)$$

Newton's law of gravitation states that every particle in the universe attracts every other particle with a force that is directly proportional to the product of the two masses and inversely proportional to the square of the distance between them.

$$F_g = G \frac{m_1 m_2}{r^2} \qquad (5.14)$$

Coulomb's law expresses the magnitude of the electrostatic force between two charged particles separated by a distance r. Opposite sign charges attract each other, and like sign charges repel.

$$F_e = k_e \frac{q_1 q_2}{r^2} \qquad (5.15)$$

SUGGESTIONS, SKILLS, AND STRATEGIES

Section 5.4 deals with the motion of a body through a gas or liquid. If you covered this section in class, the following solution to Equation 5.5 (when the resistive force $\mathbf{R} = -b\mathbf{v}$) may be useful to know:

$$\frac{dv}{dt} = g - \frac{b}{m} v \qquad (5.5)$$

In order to solve this equation, it helps to change variables. If we let $y = g - (b/m)v$, it follows that $dy = -(b/m)dv$. With these substitutions, Equation 5.5 becomes

$$-\left(\frac{m}{b}\right)\frac{dy}{dt} = y \qquad \text{or} \qquad \frac{dy}{y} = -\frac{b}{m} dt$$

Integrating this expression (now that the variables are separated) gives

$$\int \frac{dy}{y} = -\frac{b}{m} \int dt \qquad \text{or} \qquad \ln y = -\frac{b}{m} t + \text{const.}$$

This is equivalent to $y = (\text{const})\, e^{-bt/m} = g - \dfrac{b}{m} v.$ Taking $v = 0$ at $t = 0$, we see const $= g$.

so
$$v = \frac{mg}{b}(1 - e^{-bt/m}) = v_t(1 - e^{-t/\tau}) \tag{5.6}$$

where
$$\tau = m/b$$

REVIEW CHECKLIST

▷ Discuss Newton's universal law of gravity (the inverse-square law), and understand that it is an **attractive** force between two **particles** separated by a distance r.

▷ Discuss the nature of the fundamental forces in nature (gravitational, electromagnetic, weak, and nuclear) and characterize the properties and relative strengths of these forces.

▷ Apply Newton's second law to uniform and nonuniform circular motion.

▷ Recognize that motion of an object through a liquid or gas can involve resistive forces which have a complicated velocity dependence.

ANSWERS TO SELECTED CONCEPTUAL QUESTIONS

3. Suppose you are driving a car along a highway at a high speed. Why should you avoid slamming on your brakes if you want to stop in the shortest distance? That is, why should you keep the wheels turning as you brake?

Answer The brakes may lock and the car will slide farther since the coefficient of sliding friction is less than the coefficient of static friction. If the wheels continue to roll, the force of static friction will decelerate the car.

6. It has been suggested that rotating cylinders about 10 mi in length and 5 mi in diameter be placed in space and used as colonies. The purpose of the rotation is to simulate gravity for the inhabitants. Explain this concept for producing an effective gravity.

Answer The centripetal force on the inhabitants is provided by the normal force exerted on them by the cylinder wall. If the rotation rate is adjusted to such a speed that this normal force is equal to their weight on Earth, the inhabitants would not be able to distinguish between this artificial gravity and normal gravity.

□ □ □ □

13. A skydiver in free fall reaches terminal speed. After the parachute is opened, what parameters change to decrease this terminal speed?

Answer From Equation 5.7 and Newton's law, we derive the equation that governs the motion of the skydiver:

$$m\frac{dv_y}{dt} = mg - \frac{D\rho A}{2}v_y^2$$

where D is the coefficient of drag of the parachutist, and A is the area of the parachutist's body. At terminal speed,

$$a_y = dv_y/dt = 0 \quad \text{and} \quad v_t = \sqrt{2mg/D\rho A}$$

When the parachute opens, the coefficient of drag D and the effective area A both increase, thus reducing the velocity of the skydiver.

Modern parachutes also add a third term, lift, to change the equation to

$$m\frac{dv_y}{dt} = mg - \frac{D\rho A}{2}v_y^2 - \frac{L\rho A}{2}v_x^2$$

where v_y is the vertical velocity, and v_x is the horizontal velocity. This lift is best seen in the "Paraplane," an ultralight airplane made from a fan, a chair, and a parachute.

SOLUTIONS TO SELECTED END-OF-CHAPTER PROBLEMS

7. A 3.00-kg block starts from rest at the top of a 30.0° incline and slides a distance of 2.00 m down the incline in 1.50 s. Find (a) the magnitude of the acceleration of the block, (b) the coefficient of kinetic friction between block and plane, (c) the friction force acting on the block, and (d) the speed of the block after it has slid 2.00 m.

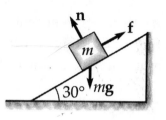

Solution We use the particle under constant acceleration and particle under net force models.

(a) At constant acceleration,
$$x_f = v_i t + \tfrac{1}{2} a t^2$$

So,
$$a = \frac{2(x_f - v_i t)}{t^2} = \frac{2(2.00 \text{ m} - 0)}{(1.50 \text{ s})^2} = 1.78 \text{ m/s}^2 \qquad \lozenge$$

From the acceleration, we can calculate the friction force, answer (c), next.

(c) Choose the x axis parallel to the incline, take the positive direction down the incline (in the direction of the acceleration) and apply the second law.

$$\Sigma F_x = mg \sin\theta - f = ma: \qquad f = m(g \sin\theta - a)$$

$$f = (3.00 \text{ kg})\left[9.80 \text{ m/s}^2(\sin 30.0°) - 1.78 \text{ m/s}^2\right] = 9.37 \text{ N} \qquad \lozenge$$

(b) Applying Newton's law in the y direction (perpendicular to the incline),

$$\Sigma F_y = n - mg \cos\theta = 0: \qquad n = mg \cos\theta$$

Because $f = \mu n$,
$$\mu = \frac{f}{mg \cos\theta} = \frac{9.37 \text{ N}}{(3.00 \text{ kg})(9.80 \text{ m/s}^2)\cos 30.0°} = 0.368 \; \lozenge$$

(d) $v_f = v_i + at$, so
$$v = 0 + (1.78 \text{ m/s}^2)(1.50 \text{ s}) = 2.67 \text{ m/s} \qquad \lozenge$$

9. Two blocks connected by a rope of negligible mass are being dragged by a horizontal force **F** (Fig. P5.9). Suppose that $F = 68.0$ N, $m_1 = 12.0$ kg, $m_2 = 18.0$ kg, and the coefficient of kinetic friction between each block and the surface is 0.100. (a) Draw a free-body diagram for each block. (b) Determine the tension, T, and the magnitude of the acceleration of the system.

Figure P5.9

Solution

(a) The free-body diagrams for m_1 and m_2 are:

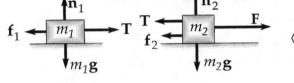

(b) Use the free-body diagrams to apply Newton's second law.

For m_1: $\qquad \Sigma F_x = T - f_1 = m_1 a \qquad$ or $\qquad T = m_1 a + f_1 \qquad$ (1)

$\qquad\qquad \Sigma F_y = n_1 - m_1 g = 0 \qquad$ or $\qquad n_1 = m_1 g$

Also, $\qquad f_1 = \mu_1 n_1 = (0.100)(12.0 \text{ kg})(9.80 \text{ m}/\text{s}^2) = 11.8$ N

For m_2: $\qquad \Sigma F_x = F - T - f_2 = m_2 a \qquad$ or $\qquad T = F - m_2 a - f_2 \qquad$ (2)

$\qquad\qquad \Sigma F_y = n_2 - m_2 g = 0 \qquad$ or $\qquad n_2 = m_2 g$

Also, $\qquad f_2 = \mu n_2 = (0.100)(18.0 \text{ kg})(9.80 \text{ m}/\text{s}^2) = 17.6$ N

Substituting T from equation (1) into (2), we get $\quad m_1 a + f_1 = F - m_2 a - f_2$

Solving for a. $\qquad a = \dfrac{F - f_1 - f_2}{m_1 + m_2} = \dfrac{(68.0 - 11.8 - 17.6) \text{ N}}{(12.0 + 18.0) \text{ kg}} = 1.29 \text{ m}/\text{s}^2$

From Eq. (1), $\qquad T = m_1 a + f_1 = (12.0 \text{ kg})(1.29 \text{ m}/\text{s}^2) + 11.8 \text{ N} = 27.2 \text{ N}$

13. A light string can support a stationary hanging load of 25.0 kg before breaking. A 3.00-kg mass attached to the light string rotates on a horizontal, frictionless table in a circle of radius 0.800 m. What range of speeds can the mass have before the string breaks?

Solution   We use the particle under net force and particle in uniform circular motion models. The string will break if the tension T exceeds

$$T_{max} = mg = (25.0 \text{ kg})(9.80 \text{ m/s}^2) = 245 \text{ N}$$

r = 0.800 m

As the 3.00-kg mass rotates in a horizontal circle, the tension provides the centripetal acceleration:

$$a = v^2 / r$$

From $\Sigma F = ma$, $$T = mv^2 / r$$

Therefore, $$v^2 = \frac{rT}{m} = \frac{(0.800 \text{ m})T}{(3.00 \text{ kg})} \le \frac{(0.800 \text{ m})}{(3.00 \text{ kg})} T_{max}$$

Substituting $T_{max} = 245$ N, we find $v^2 \le 65.3 \text{ m}^2 / \text{s}^2$ and $0 < v < 8.08 \text{ m/s}$ ◊

---

15. A crate of eggs is located in the middle of the flatbed of a pickup truck as the truck negotiates an unbanked curve in the road. The curve may be regarded as an arc of a circle of radius 35.0 m. If the coefficient of static friction between crate and truck is 0.600, how fast can the truck be moving without the crate sliding?

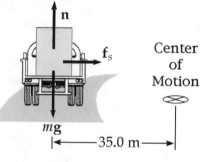

Solution

Call the mass of the egg crate m. The forces on it are its weight mg vertically down, the normal force $\mathbf{n}$ of the truck bed vertically up, and static friction $\mathbf{f}_s$ directed to oppose relative sliding motion of the crate over the truck bed. The friction force is directed radially inward. It is the only horizontal force on the crate, so it must provide the centripetal acceleration. When the truck has maximum speed, friction f_s will have its maximum value with $f_s = \mu_s n$.

$\Sigma F_y = ma_y$ gives $\qquad\qquad n - mg = 0 \qquad$ or $\qquad n = mg$

$\Sigma F_x = ma_x$ gives $\qquad\qquad f_s = ma_r$

From these two equations, $\qquad\qquad \mu_s n = \dfrac{mv^2}{r} \qquad$ and $\qquad \mu_s mg = \dfrac{mv^2}{r}$

The mass divides out, leaving $\qquad v = \sqrt{\mu_s gr} = \sqrt{(0.600)(9.80 \text{ m}/\text{s}^2)(35.0 \text{ m})} = 14.3 \text{ m}/\text{s}$ ◊

---

17. Tarzan (m = 85.0 kg) tries to cross a river by swinging from a vine. The vine is 10.0 m long, and his speed at the bottom of the swing (as he just clears the water) is 8.00 m/s. Tarzan doesn't know that the vine has a breaking strength of 1000 N. Does he make it safely across the river?

Solution   The forces acting on Tarzan are the force of gravity $m\mathbf{g}$ and the force from the rope, **T**. At the lowest point in his motion, **T** is upward and $m\mathbf{g}$ is downward as in the free-body diagram. Thus, Newton's second law gives

$$T - mg = \frac{mv^2}{r}$$

Solving for T, with v = 8.00 m/s, r = 10.0 m, and m = 85.0 kg, gives

$$T = m\left(g + \frac{v^2}{r}\right) = (85.0 \text{ kg})\left(9.80 \text{ m}/\text{s}^2 + \frac{(8.00 \text{ m}/\text{s})^2}{10.0 \text{ m}}\right) = 1.38 \times 10^3 \text{ N}$$

Since T **exceeds** the breaking strength of the vine (1000 N), Tarzan **doesn't make it!** The vine breaks **before** he reaches the bottom of the swing. ◊

---

19. A pail of water is rotated in a vertical circle of radius 1.00 m. What is the minimum speed of the pail when it is upside down at the top of the circle if no water is to spill out?

Solution   The normal force, **n**, will maintain exactly enough force to prevent the water from going **through** the bottom of the bucket. If water were to spill out, the force of gravity would exceed that required to provide the centripetal acceleration, and the normal force would be zero:

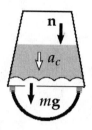

$$ma_c < mg \qquad \text{or} \qquad \frac{mv^2}{r} < mg$$

Since that is the only case in which water would spill out, all other cases must result in no water spilling out. That is,

$$\frac{mv^2}{r} \geq mg \qquad \text{or} \qquad v^2 \geq rg$$

At the minimum speed, we have $v_{min} = \sqrt{rg} = \sqrt{(1.00 \text{ m})(9.80 \text{ m} / \text{s}^2)} = 3.13 \text{ m} / \text{s}$ ◊

---

25. A motor boat cuts its engine when its speed is 10.0 m/s and coasts to rest. The equation describing the motion of the motorboat during this period is $v = v_i e^{-ct}$, where v is the speed at time t, v_i is the initial speed, and c is a constant. At $t = 20.0$ s, the speed is 5.00 m/s. (a) Find the constant c. (b) What is the speed at $t = 40.0$ s? (c) Differentiate the expression for $v(t)$ and thus show that the acceleration of the boat is proportional to the speed at any time.

Solution

(a)   We must fit the equation $v = v_i e^{-ct}$ to the two data points:

At $t = 0$, $v = 10.0$ m/s:

$$v = v_i e^{-ct}$$

$$10.0 \text{ m/s} = v_i e^0 = v_i \times 1$$

$$v_i = 10.0 \text{ m/s}$$

At $t = 20.0$ s, $v = 5.00$ m/s: 5.00 m/s $= (10.0$ m/s$)\,e^{-c(20.0\text{ s})}$

$$0.500 = e^{-c(20.0\text{ s})}$$

$$\ln 0.500 = (-c)(20.0\text{ s})$$

$$c = \frac{-\ln 0.500}{20.0\text{ s}} = 0.0347\text{ s}^{-1} \qquad \Diamond$$

(b) At all times $v = (10.0$ m/s$)\,e^{-0.0347t}$

At $t = 40.0$ s, $v = (10.0$ m/s$)\,e^{-0.0347 \times 40.0} = 2.50$ m/s $\Diamond$

(c) The acceleration is the rate of change of the velocity:

$$a = \frac{dv}{dt} = \frac{d}{dt}v_i e^{-ct} = v_i\left(e^{-ct}\right)(-c) = -c\left(v_i e^{-ct}\right) = -cv = \left(-0.0347\text{ s}^{-1}\right)v$$

Thus, the acceleration is a negative constant times the speed. $\Diamond$

---

27. A hailstone of mass 4.80×10^{-4} kg falls through the air and experiences a net force given by $F = -mg + Cv^2$ where $C = 2.50 \times 10^{-5}$ kg/m. (a) Calculate the terminal speed of the hailstone. (b) Use Euler's method of numerical analysis to find the speed and position of the hailstone at 0.2-s intervals, taking the initial speed to be zero. Continue the calculation until the hailstone reaches 99% of terminal speed.

Solution

(a) At terminal speed the acceleration and total force are zero: $-mg + Cv^2 = 0$
We choose the negative root, because the velocity is downward.

$$v = \pm\sqrt{\frac{mg}{C}} = \pm\sqrt{\frac{\left(4.80 \times 10^{-4}\text{ kg}\right)\left(9.80\text{ m/s}^2\right)}{2.50 \times 10^{-5}\text{ kg/m}}} = -13.7\text{ m/s} \qquad \Diamond$$

(b) At $t = 0$, we take $x = 0$. The hailstone starts from rest, so $v = 0$; we then cycle through the following set of equations:

$$a = \frac{\Sigma F}{m} = \frac{-mg + Cv^2}{m} = -g + \frac{Cv^2}{m} \tag{1}$$

$$x_{new} = x_{old} + v_{old}\Delta t \tag{2}$$

$$v_{new} = v_{old} + a_{old}\Delta t \tag{3}$$

and we loop back to Equation (1) to find the next acceleration. The values we find are listed below, first for every step, and then (after 2.0 s), for every fifth step:

t (s)	0.00	0.200	0.400	0.600	0.800	1.00	1.20
a (m/s²)	−9.80	−9.60	−9.02	−8.12	−7.02	−5.85	−4.72
x (m)	0.00	0.00	−0.392	−1.17	−2.30	−3.77	−5.51
v (m/s)	0.00	−1.96	−3.88	−5.68	−7.31	−8.71	−9.88

t (s)	1.40	1.60	1.80	2.00	3.00	4.00	5.00
a (m/s²)	−3.70	−2.84	−2.14	−1.59	−0.321	−.0606	−0.0113
x (m)	−7.48	−9.65	−12.0	−14.4	−27.4	−41.0	−54.7
v (m/s)	−10.8	−11.6	−12.1	−12.6	−13.5	−13.7	−13.7

The hailstone never attains terminal speed exactly, but it passes 99% of it at 3.4 seconds, and 99.99% at 6.2 seconds.

◊

35. When a falling meteor is at a distance above the Earth's surface of 3.00 times the Earth's radius, what is its free-fall acceleration due to the gravitational force exerted on it?

Solution   The acceleration of gravity on Earth, $g = Gm/r^2$, follows an inverse-square law. At the surface (at distance one Earth-radius R_E from the center), it is 9.80 m/s².

At an altitude $3R_E$ above the surface (at distance $4R_E$ from the center), the acceleration of gravity will be $4^2 = 16$ times smaller:

$$g = \frac{GM_E}{(4R_E)^2} = \frac{GM_E}{16R_E^2} = \frac{9.80 \text{ m/s}^2}{16} = 0.613 \text{ m/s}^2 \text{ down}$$

◊

49. Because the Earth rotates about its axis, a point on the equator experiences a centripetal acceleration of 0.0337 m/s², but a point at the poles experiences no centripetal acceleration. (a) Show that at the Equator the magnitude of the gravitational force on an object must exceed the normal force required to the support the object. That is, show that the object's true weight exceeds its apparent weight. (b) What is the apparent weight at the Equator and at the poles of a person having a mass of 75.0 kg? (Assume the Earth is a uniform sphere and take $g = 9.800$ N/kg.)

Solution

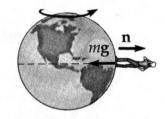

Let **n** represent the force exerted on the object by the scale, which is the "apparent weight." The true weight is $m\mathbf{g}$. The centripetal acceleration toward the center of the Earth is

$$a_c = \frac{v^2}{R_E}$$

(a) Summing up forces on the object in the direction towards the Earth's center gives:

(1) $\quad mg - n = ma_c$

Thus, we see that

$n = m(g - a_c) < mg$

or

(2) $\quad mg = n + ma_c > n$ ◊

(b) Taking $m = 75.0$ kg, $a_c = 0.0337$ m/s², and $g = 9.800$ m/s²,

at the Equator: $\quad n = m(g - a_c) = (75.0 \text{ kg})(9.800 \text{ m/s}^2 - 0.0337 \text{ m/s}^2) = 732$ N ◊

at the Poles: $\quad n = mg = (75.0 \text{ kg})(9.800 \text{ m/s}^2) = 735$ N $\quad (a_c = 0)$ ◊

53. An amusement park ride consists of a large vertical cylinder that spins about its axis fast enough that any person inside is held up against the wall when the floor drops away (Fig. P5.53). The coefficient of static friction between a person and the wall is μ_s, and the radius of the cylinder is R. (a) Show that the maximum period of revolution necessary to keep the person from falling is given by $T = (4\pi^2 R \mu_s / g)^{1/2}$. (b) Obtain a numerical value for T if $R = 4.00$ m and $\mu_s = 0.400$. How many revolutions per minute does the cylinder make?

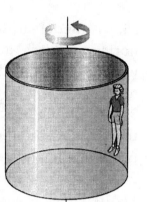

Figure P5.53

Solution   We model the person as a particle in uniform circular motion.

(a)   The wall's normal force pushes inward: $n = \dfrac{mv^2}{R} = \dfrac{m}{R}\left(\dfrac{2\pi R}{T}\right)^2 = \dfrac{4\pi^2 R m}{T^2}$

The friction and weight balance: $f_s = \mu_s n = mg$

Therefore, with $\mu_s n = mg$, $\mu_s n = \mu_s \dfrac{4\pi^2 R m}{T^2} = mg$,

Solving, $T^2 = \dfrac{4\pi^2 R \mu_s}{g}$ gives $T = \sqrt{\dfrac{4\pi^2 R \mu_s}{g}}$ ◊

(b) $T = \sqrt{\dfrac{4\pi^2 (4.00 \text{ m})(0.400)}{9.80 \text{ m} / \text{s}^2}} = 2.54$ s   ◊

The angular speed is $\left(\dfrac{1 \text{ rev}}{2.54 \text{ s}}\right)\left(\dfrac{60 \text{ s}}{\text{min}}\right) = 23.6$ rev / min   ◊

Related Questions and Answers: Why is the normal force horizontally inward? Because the wall is vertical, and on the outside.
Why is there no upward normal force? Because there is no floor.
Why is the frictional force directed upward? The friction opposes the possible relative motion of the person sliding down the wall.
Why is it not kinetic friction? Because person and wall are moving together, stationary with respect to each other.
Why is there no outward force on her? No other object pushes out on her. She pushes out on the wall as the wall pushes inward on her.

55. A model airplane of mass 0.750 kg flies in a horizontal circle at the end of a 60.0-m control wire, with a speed of 35.0 m/s. Compute the tension in the wire if it makes a constant angle of 20.0° with the horizontal. The forces exerted on the airplane are the pull of the control wire, its own weight, and aerodynamic lift, which acts at 20.0° inward from the vertical as shown in Figure P5.55.

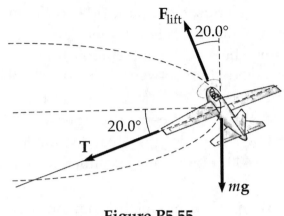

Figure P5.55

Solution

The plane's acceleration is toward the center of the circle of motion, so it is horizontal. The radius of the circle of motion is (60.0 m) cos 20.0° = 56.4 m, and the acceleration is

$$a_c = \frac{v^2}{r} = \frac{(35.0 \text{ m}/\text{s})^2}{56.4 \text{ m}} = 21.7 \text{ m}/\text{s}^2$$

We can also calculate the weight of the airplane:

$$F_g = mg = (0.750 \text{ kg})(9.80 \text{ m}/\text{s}^2) = 7.35 \text{ N}$$

We define our axes for convenience. In this case, two of the forces—one of them our force of interest—are directed along the 20.0° lines.

We define the x-axis to be directed in the (+**T**) direction, and the y-axis to be directed in the direction of lift. With these definitions, the x component of the centripetal acceleration is

$$a_{cx} = a_c \cos(20.0°)$$

and $\Sigma F_x = ma_x$ yields

$$T + F_g \sin(20.0°) = ma_{cx}$$

Solving for T,

$$T = ma_{cx} - F_g \sin(20.0°)$$

$$T = (0.750 \text{ kg})(21.7 \text{ m}/\text{s}^2)(\cos 20.0°) - (7.35 \text{ N})\sin 20.0°$$

and

$$T = (15.3 \text{ N}) - (2.51 \text{ N}) = 12.8 \text{ N} \qquad \Diamond$$

Chapter 6

Energy and Energy Transfer

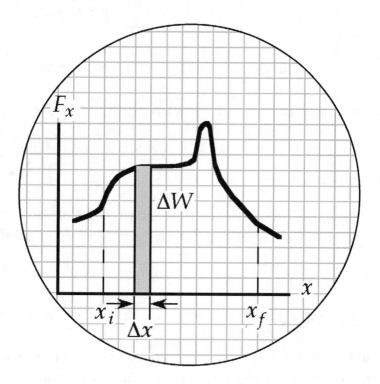

INTRODUCTION

The transformation of energy from one form to another is an essential part of the study of physics, engineering, chemistry, biology, geology, and astronomy. When energy is changed from one form to another, the total amount of energy remains the same. Conservation of energy says that if an isolated system loses energy in some form, it gains an equal amount of energy in one or more other forms.

In this chapter, we introduce the concepts of work and energy. Work is done by a force acting on an object when the point of application of that force moves through some distance and the force has a component along the line of motion. Kinetic energy is energy associated with the motion of an object. The concepts of work and energy can be applied to the dynamics of a mechanical system without resorting to Newton's laws. However, it is important to note that the work-energy concepts are based upon Newton's laws and therefore do not involve any new physical principles. In a complex situation, the "energy approach" can often provide a much simpler analysis than the direct application of Newton's second law.

NOTES FROM SELECTED CHAPTER SECTIONS

6.1 Systems and Environments

A very important step in solving problems with an energy approach is to correctly identify the system. An isolated system does not interact with its environment. A nonisolated system is one that does interact with the environment, so that there is an influence from the environment across the system boundary.

6.2 Work Done By a Constant Force

Work done by a constant force is defined as the product of the component of the force in the direction of the displacement and the magnitude of the displacement.

The **unit of work** in the SI system is the newton·meter, N·m: 1 newton·meter = 1 joule (J).

6.3 The Scalar Product of Two Vectors

The **scalar product** or dot product of any two vectors is a scalar quantity equal to the product of the magnitudes of the two vectors and the cosine of the angle included between the directions of the two vectors.

6.4 Work Done By a Varying Force

Work done by a varying force is equal to the area under the force-displacement curve.

6.5 Kinetic Energy and the Work-Kinetic Energy Theorem

If a system interacts with its environment such that a force acting on the system moves through a displacement and the only type of change in the system is in its kinetic energy, the work done by the force on the system is equal to the change in kinetic energy of the system. This is known as the **work-kinetic energy theorem**.

6.8 Power

Power is the time rate of doing work or transferring energy. The SI unit of power is the watt, W. 1 W = 1 J/s.

EQUATIONS AND CONCEPTS

The work done by an agent exerting a **constant** force **F** on a system is defined to be the product of the displacement of the point of application of the force and the component of force in the direction of the displacement. Work is a scalar quantity.

$$W \equiv F\Delta r \cos\theta \qquad (6.1)$$

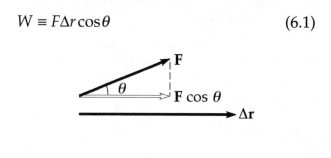

The dot product of any two vectors **A** and **B** is defined to be a scalar quantity whose magnitude is AB cos θ. (The product of the magnitudes of the two vectors and the cosine of the smaller of the two angles between **A** and **B**.)

$$\mathbf{A} \cdot \mathbf{B} \equiv AB \cos\theta \qquad (6.3)$$

It is convenient to express the work done by a constant force as the **dot product** (scalar product) of the force and displacement vectors.

$$W = \mathbf{F} \cdot \Delta\mathbf{r} = F\Delta r \cos\theta \qquad (6.4)$$

Note that work done by a force can be positive, negative, or zero depending on the value of θ. Work done by a force is positive if **F** has a component in the direction of $\Delta\mathbf{r}$ $(0 \le \theta < 90°)$; W is negative if the projection of **F** onto $\Delta\mathbf{r}$ is opposite to $\Delta\mathbf{r}$ $(90° < \theta < 180°)$. Finally, W is zero if **F** is perpendicular to $\Delta\mathbf{r}$ $(\theta = \pm 90°)$. Work is a **scalar** quantity which has SI units of joules (J), where $1\,\text{J} = 1\,\text{N·m}$.

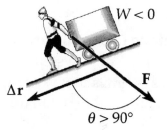

The scalar product of two vectors **A** and **B** can be expressed in terms of the $x, y,$ and z components of the two vectors.

$$\mathbf{A} \cdot \mathbf{B} = A_x B_x + A_y B_y + A_z B_z \qquad (6.9)$$

If a force acting along x varies with position, and the body is displaced from x_i to x_f, the **work done by that force** is given by an integral expression. Graphically, the work done equals the area under the F_x versus x curve.

$$W = \int_{x_i}^{x_f} F_x\, dx \qquad (6.11)$$

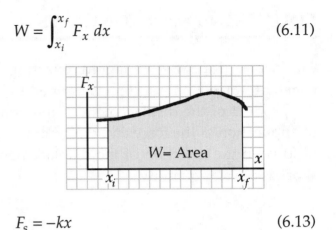

The force exerted by a stretched or compressed spring is directed opposite the displacement. The force law for springs (Equation 6.13) is known as Hooke's law.

$$F_s = -kx \qquad (6.13)$$

If a mass is connected to a spring of force constant k, **the work done by the spring force** $(-kx)$ as the mass undergoes an arbitrary displacement from x_i to x_f is given by Equation 6.15.

$$W_s = \int_{x_i}^{x_f} (-kx)dx = \tfrac{1}{2}kx_i^2 - \tfrac{1}{2}kx_f^2 \qquad (6.15)$$

Kinetic energy, K, is energy associated with the motion of an object.

$$K \equiv \tfrac{1}{2}mv^2 \qquad (6.18)$$

The **work-kinetic energy theorem** states that when work is done on a system and the only change in the system is in its speed, the work done by the net force is equal to the change in kinetic energy of the system.

$$W_{net} = K_f - K_i = \Delta K \qquad (6.19)$$

$$W_{net} = \tfrac{1}{2}mv_f^2 - \tfrac{1}{2}mv_i^2 \qquad (6.17)$$

Note that if W_{net} is positive, the kinetic energy increases; if W_{net} is negative, the kinetic energy decreases. Therefore, the speed of a body will only change if there is net work done on it. (When $W_{net} = 0$, the resultant work is zero, and $\Delta K = 0$.)

$$\tfrac{1}{2}mv_i^2 + W_{net} = \tfrac{1}{2}mv_f^2$$

As we will later study in detail, the total energy E of a system includes the kinetic energy and also other forms of energy. Energy H can cross the boundary of a system by work and also by other methods. The principle of **conservation of energy** is expressed by the **continuity equation for energy**.

$$\Delta E_{system} = \Sigma H \tag{6.20}$$

The work-kinetic energy theorem takes a special form if friction is among the forces acting on an object.

$$K_i - f_k \Delta x + \Sigma W_{other\ forces} = K_f \tag{6.21}$$

A friction force transforms kinetic energy to internal energy, and the increase in internal energy is equal to the decrease in kinetic energy.

$$\Delta E_{int} = f_k \Delta x \tag{6.24}$$

The **average power** supplied by a force is the ratio of the work done by that force to the time interval over which it acts.

$$\overline{\mathcal{P}} \equiv \frac{W}{\Delta t} \tag{6.25}$$

The **instantaneous power** is equal to the limit of the average power as the time interval approaches zero.

$$\mathcal{P} = \frac{dW}{dt} = \mathbf{F} \cdot \frac{d\mathbf{r}}{dt} = \mathbf{F} \cdot \mathbf{v} \tag{6.27}$$

The SI unit of power is J/s, which is called a watt (W). The kilowatt-hour is a unit of energy.

$$1\ W = 1\ J/s = 1 kg \cdot m^2/s^2$$

$$1\ kWh = 3.60 \times 10^6\ J$$

The unit of power in the British engineering system is the horsepower.

$$1\ hp = 746\ W$$

SUGGESTIONS, SKILLS, AND STRATEGIES

There are two new mathematical skills you must learn. The first is the definition of the scalar (or dot) product, $\mathbf{A} \cdot \mathbf{B} \equiv AB\cos\theta$, where θ is the angle between $\mathbf{A}$ and $\mathbf{B}$. Since $\mathbf{A} \cdot \mathbf{B}$ is a scalar, the order of the terms can be interchanged. That is, $\mathbf{A} \cdot \mathbf{B} = \mathbf{B} \cdot \mathbf{A}$. Furthermore, $\mathbf{A} \cdot \mathbf{B}$ can be positive, negative, or zero depending on the value of θ. (That is, $\cos\theta$ varies from -1 to $+1$.) If vectors are expressed in unit vector form, then the dot product is conveniently carried out using the multiplication table for unit vectors:

$$\mathbf{i} \cdot \mathbf{i} = \mathbf{j} \cdot \mathbf{j} = \mathbf{k} \cdot \mathbf{k} = 1; \qquad \mathbf{i} \cdot \mathbf{j} = \mathbf{i} \cdot \mathbf{k} = \mathbf{j} \cdot \mathbf{k} = 0$$

The second operation introduced in this chapter is the definite integral. In Section 6.3, it is shown that the work done by a **variable** force F_x in displacing a particle a small distance Δx is given by

$$\Delta W \approx F_x \, \Delta x$$

(ΔW equals the area of the shaded rectangle in Figure 6.1.) The total work done by F_x as the particle is displaced from x_i to x_f is given approximately by the **sum** of such terms.

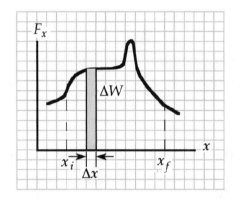

Figure 6.1

If we take such a sum, letting the widths of the displacements approach dx, the number of terms in the sum becomes very large and we get the actual work done:

$$W = \lim_{\Delta x \to 0} \sum_{x_i}^{x_f} F_x \, \Delta x = \int_{x_i}^{x_f} F_x \, dx$$

The quantity on the right is a definite integral, which graphically represents the **area under the F_x versus x curve**, as in Figure 6.1.

REVIEW CHECKLIST

▷ Define the work done by a constant force, and realize that work is a scalar. Describe the work done by a force which **varies** with position. In the one-dimensional case, note that the work done equals the area under the F_x versus x curve.

▷ Take the scalar or dot product of any two vectors **A** and **B** using the definition $\mathbf{A} \cdot \mathbf{B} \equiv AB\cos\theta$, or by writing **A** and **B** in unit vector form and using the multiplication table for unit vectors.

▷ Define the kinetic energy of an object of mass m moving with a speed v.

Relate the work done by the net force on an object to the **change** in kinetic energy. The relation $W_{net} = \Delta K = K_f - K_i$ is called the work-kinetic energy theorem, and is valid whether or not the (resultant) force is constant. That is, if we know the net work done on a particle as it undergoes a displacement, we also know the **change** in its kinetic energy.

▷ Energy is conserved. If the amount of energy in a system changes, the change can only be due to a transfer of energy across the boundary of the system. This is the most important concept in this chapter, so you must understand it thoroughly.

▷ Define the concepts of average power and instantaneous power (the time rate of doing work).

ANSWERS TO SELECTED CONCEPTUAL QUESTIONS

6. As a simple pendulum swings back and forth, the forces acting on the suspended mass are the gravitational force, the tension in the supporting cord, and air resistance. (a) Which of these forces, if any, does no work on the pendulum? (b) Which of these forces does negative work at all times during its motion? (c) Describe the work done by the gravitational force while the pendulum is swinging.

Answer   (a) The tension in the supporting cord does no work, because the motion of the pendulum is always perpendicular to the cord, and therefore to the force exerted by the string. (b) The air resistance does negative work at all times, since the air resistance is always acting in a direction opposite to the motion. (c) The weight always acts downward; therefore, the work done by the gravitational force is positive on the downswing, and negative on the upswing.

□   □   □   □

9. Can kinetic energy be negative? Explain.

Answer   No. Kinetic energy $= mv^2/2$. Since v^2 is always positive, K is always positive.

□   □   □   □

11. If the speed of a particle is doubled, what happens to its kinetic energy?

Answer   Kinetic energy, K, depends on the square of the velocity. Therefore if the speed is doubled, the kinetic energy will increase by a factor of four.

□   □   □   □

SOLUTIONS TO SELECTED END-OF-CHAPTER PROBLEMS

1. A block of mass 2.50 kg is pushed 2.20 m along a frictionless horizontal table by a constant 16.0-N force directed 25.0° below the horizontal. Determine the work done by (a) the applied force, (b) the normal force exerted by the table, (c) the force of gravity. (d) Determine the total work done on the block.

Solution $W = F\Delta r \cos\theta$

(a) By the applied force, $\qquad W_{app} = (16.0 \text{ N})(2.20 \text{ m})\cos(25.0°) = 31.9 \text{ J}$ ◊

(b) By normal force, $\qquad W_n = n\Delta r \cos\theta = n\Delta r \cos(90°) = 0$ ◊

(c) By force of gravity, $\qquad W_g = F_g \Delta r \cos\theta = mg\Delta r \cos(90°) = 0$ ◊

(d) Net work done on the block: $\quad W_{net} = W_{app} + W_n + W_g = 31.9 \text{ J}$ ◊

3. A certain superhero, whose mass is 80.0 kg, is dangling on to the free end of a 12.0-m rope, the other end of which is fixed to a tree limb above. He is able to get the rope in motion as only he knows how, eventually getting it to swing enough that he can reach a ledge when the rope makes a 60.0° angle with the vertical. How much work was done by the gravitational force on the superhero in this maneuver?

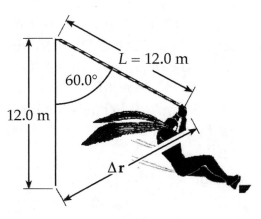

Solution   The work done is $W = \mathbf{F} \cdot \Delta\mathbf{r}$, where

The gravitational force is $F_g = -mg\,\mathbf{j} = (80.0 \text{ kg})(-9.80 \text{ m/s}^2\,\mathbf{j}) = -784\,\mathbf{j}$ N

The superhero travels $\Delta\mathbf{r} = L\sin 60°\,\mathbf{i} + L\cos 60°\,\mathbf{j} = (12.0 \text{ m})\sin 60°\,\mathbf{i} + (12.0 \text{ m})\cos 60°\,\mathbf{j}$

Thus, $W = \mathbf{F} \cdot \Delta\mathbf{r} = (-784\,\mathbf{j} \text{ N}) \cdot (10.39\,\mathbf{i} \text{ m} + 6.00\,\mathbf{j} \text{ m}) = -4.70 \times 10^3$ J   ◊

---

7. A force $\mathbf{F} = (6\mathbf{i} - 2\mathbf{j})$ N acts on a particle that undergoes a displacement $\Delta\mathbf{r} = (3\mathbf{i} + \mathbf{j})$ m. Find (a) the work done by the force on the particle and (b) the angle between $\mathbf{F}$ and $\Delta\mathbf{r}$.

Solution   We use the mathematical representation of the definition of work.

(a) $W = \mathbf{F} \cdot \Delta\mathbf{r} = (6\,\mathbf{i} - 2\,\mathbf{j}) \cdot (3\,\mathbf{i} + 1\,\mathbf{j}) = (6 \text{ N})(3 \text{ m}) + (-2 \text{ N})(1 \text{ m}) = 18 \text{ J} - 2 \text{ J} = 16 \text{ J}$ ◊

(b) $|\mathbf{F}| = \sqrt{F_x^2 + F_y^2} = \sqrt{6^2 + (-2)^2}$ N $= 6.32$ N

$|\Delta\mathbf{r}| = \sqrt{\Delta r_x^2 + \Delta r_y^2} = \sqrt{3^2 + 1^2}$ m $= 3.16$ m

$W = F\Delta r \cos\theta$: $\cos\theta = \dfrac{W}{F\,\Delta r} = \dfrac{16.0 \text{ J}}{(6.32 \text{ N})(3.16 \text{ m})} = 0.800$

and $\theta = \cos^{-1}(0.800) = 36.9°$ ◊

101

11. A particle is subject to a force F_x that varies with position as in Figure P6.11. Find the work done by the force on the object as it moves (a) from $x = 0$ to $x = 5.00$ m, (b) from $x = 5.00$ m to $x = 10.0$ m, and (c) from $x = 10.0$ m to $x = 15.0$ m. (d) What is the total work done by the force over the distance $x = 0$ to $x = 15.0$ m?

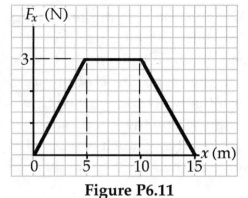

Figure P6.11

Solution $W = \int F_x dx$.

We use the graphical representation of the definition of work. W equals the area under the force-displacement curve.

(a) For the region $0 \le x \le 5.00$ m, $\quad W = \dfrac{(3.00 \text{ N}) (5.00 \text{ m})}{2} = 7.50 \text{ J}$ ◊

(b) For the region $5.00 \text{ m} \le x \le 10.0$ m, $\quad W = (3.00 \text{ N}) (5.00 \text{ m}) = 15.0 \text{ J}$ ◊

(c) For the region $10.0 \text{ m} \le x \le 15.0$ m, $\quad W = \dfrac{(3.00 \text{ N}) (5.00 \text{ m})}{2} = 7.50 \text{ J}$ ◊

(d) For the region $0 \le x \le 15.0$ m, $\quad W = (7.50 + 7.50 + 15.0) \text{ J} = 30.0 \text{ J}$ ◊

---

23. A 2100-kg pile driver is used to drive a steel I-beam into the ground. The pile driver falls 5.00 m before coming into contact with the top of the beam, and it drives the beam 12.0 cm farther into the ground before coming to rest. Using energy considerations, calculate the average force the beam exerts on the pile driver while the pile driver is brought to rest.

Solution We use the energy version of the nonisolated system model. Choose the initial point when the mass is elevated and the final point when it comes to rest again 5.12 m below. Two forces, gravity and the normal force exerted by the beam on the pile driver, do work on the pile driver:

$K_i + W_{net} = K_f$: $\qquad 0 + mg\Delta r_w \cos 0° + n\Delta r_n \cos 180° = 0$

where $\qquad \Delta r_w = 5.12$ m and $\Delta r_n = 0.120$ m, and $m = 2100$ kg

In this situation, the weight vector is in the direction of motion and the beam exerts a force on the piledriver which is opposite the direction of motion.

$$(2100 \text{ kg})(9.80 \text{ m/s}^2)(5.12 \text{ m}) + n(0.120 \text{ m})(-1) = 0$$

Solve for n: $$n = \frac{1.05 \times 10^5 \text{ J}}{0.120 \text{ m}} = 878 \text{ kN}$$ ◊

Additional Calculation: Show that the work done by gravity on an object can be represented by mgh, where h is the vertical height that the object falls. Apply your results to the problem above.

By the figure to the right, where Δr is the path of the object,

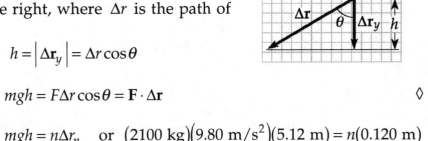

$$h = \left| \Delta \mathbf{r}_y \right| = \Delta r \cos \theta$$

Since $F = mg$, $mgh = F \Delta r \cos \theta = \mathbf{F} \cdot \Delta \mathbf{r}$ ◊

In this problem, $mgh = n \Delta r_n$ or $(2100 \text{ kg})(9.80 \text{ m/s}^2)(5.12 \text{ m}) = n(0.120 \text{ m})$

and $n = 878 \text{ kN}$ ◊

---

29. A 40.0-kg box initially at rest is pushed 5.00 m along a rough, horizontal floor with a constant applied horizontal force of 130 N. If the coefficient of friction between the box and the floor is 0.300, find (a) the work done by the applied force, (b) the increase in internal energy due to friction, (c) the work done by the normal force (d) the work done by gravity, (e) the change in kinetic energy of the box, and (f) the final speed of the box.

Solution $\mu_k = 0.300$
$\Delta r = 5.00 \text{ m}$
$v_i = 0$

(a) The applied force and the motion are both **horizontal**.

$$W_F = \mathbf{F} \cdot \Delta \mathbf{r} = F \Delta r \cos(0°) = (130 \text{ N})(5.00 \text{ m})(1) = 650 \text{ J}$$ ◊

(b) $f_k = \mu_k n = \mu_k mg = 0.300(40.0 \text{ kg})(9.80 \text{ m}/\text{s}^2) = 117.6 \text{ N}$

$$\Delta E_{\text{int}} = f_k \Delta x = (117.6 \text{ N})(5.00 \text{ m}) = 588 \text{ J}$$ ◊

(c) Since the normal force is perpendicular to the motion,

$$W_n = F\Delta r \cos(90°) = (130 \text{ N})(5.00 \text{ m})(0) = 0$$ ◊

(d) The force of gravity is also perpendicular to the motion, so $W_g = 0$ ◊

We use the energy version of the nonisolated system model.

(e) $\Delta K = W_{\text{other forces}} - f_k \Delta r = 650 \text{ J} - 588 \text{ J} = 62.0 \text{ J}$ ◊

(f) $\frac{1}{2}mv_i^2 + W_{\text{other forces}} - f_k \Delta r = \frac{1}{2}mv_f^2$

$$v_f = \sqrt{\frac{2}{m}\left[\Delta K + \frac{1}{2}mv_i^2\right]} = \sqrt{\left(\frac{2}{40.0 \text{ kg}}\right)\left[62.0 \text{ J} + \frac{1}{2}(40.0 \text{ kg})(0)^2\right]} = 1.76 \text{ m/s}$$ ◊

---

31. A sled of mass m is given a kick on a frozen pond. The kick imparts to it an initial speed of 2.00 m/s. The coefficient of kinetic friction between sled and ice is 0.100. Use energy considerations to find the distance the sled moves before it stops.

Solution Using Newton's law in the vertical direction, $\Sigma F_y = ma_y$: $+n - mg = 0$,

In addition, we know from the law of friction $f_k = \mu_k n = \mu_k mg$

In applying the work-kinetic energy theorem, we do not consider the kick, but take the original point after the kick, with $v_i = 2.00$ m/s, and the final point where the sled stops moving. The weight and normal force, each at 90° to the direction of motion, do no work.

$$K_i + W_{\text{other forces}} - f_k \Delta r = K_f : \qquad \frac{1}{2}mv_i^2 - f_k \Delta r = 0$$

Solving, $$\Delta r = \frac{mv_i^2}{2f_k} = \frac{mv_i^2}{2\mu_k mg} = \frac{v_i^2}{2\mu_k g} = \frac{(2.00 \text{ m/s})^2}{2(0.100)(9.80 \text{ m/s}^2)} = 2.04 \text{ m}$$ ◊

---

of constant

2 m/s

70

33. A crate of mass 10.0 kg is pulled up a rough incline with an initial speed of 1.50 m/s. The pulling force is 100 N parallel to the incline, which makes an angle of 20.0° with the horizontal. The coefficient of kinetic friction is 0.400, and the crate is pulled 5.00 m. (a) How much work is done by gravity? (b) Determine the increase in internal energy due to friction (c) How much work is done by the 100-N force? (d) What is the change in kinetic energy of the crate? (e) What is the speed of the crate after being pulled 5.00 m?

Solution The force of gravity is $(10.0 \text{ kg})(9.80 \text{ m/s}^2) = 98.0$ N straight down, at an angle of $(90.0° + 20.0°) = 110.0°$ with the motion. The work done by gravity on the crate is

(a) $W_g = \mathbf{F} \cdot \Delta\mathbf{r} = (98.0 \text{ N})(5.00 \text{ m}) \cos 110.0° = -168$ J ◊

(b) Setting the x-y axes parallel and perpendicular to the incline,

From $\Sigma F_y = ma_y$, $+n - (98.0 \text{ N})\cos 20.0° = 0$ $n = 92.1$ N

and $f_k = \mu_k n = 0.400 (92.1 \text{ N}) = 36.8$ N

Therefore, $\Delta E_{\text{int}} = f_k \Delta x = (36.8 \text{ N})(5.00 \text{ m}) = 184$ J ◊

(c) $W = \mathbf{F} \cdot \Delta\mathbf{r} = 100 \text{ N} (5.00 \text{ m}) \cos 0° = +500$ J ◊

(d) We use the energy version of the nonisolated system model.

By Equation 6.23, $\Delta K = -f_k \Delta x + \Sigma W_{\text{other forces}}$

$\Delta K = -f_k \Delta x + W_g + W_{\text{applied force}} + W_n = -168 \text{ J} - 184 \text{ J} + 500 \text{ J} + 0 = 148$ J ◊

The normal force does zero work, because it is at 90° to the motion.

(e) Since $K_f = K_i + \Delta K$, $\frac{1}{2}(10.0 \text{ kg}) v_f^2 = \frac{1}{2}(10.0 \text{ kg})(1.50 \text{ m/s})^2 + 148 \text{ J} = 159$ J

Thus, $v_f = \sqrt{\dfrac{2(159 \text{ kg} \cdot \text{m}^2 / \text{s}^2)}{10.0 \text{ kg}}} = 5.65$ m/s ◊

35. A 700-N Marine in basic training climbs a 10.0-m vertical rope at a constant speed in 8.00 s. What is his power output?

Solution   The marine must exert a 700 N upward force opposite the gravitational force to lift his body at constant speed. Then his muscles do work:

$$W = \mathbf{F} \cdot \Delta \mathbf{r} = (700\mathbf{j} \text{ N}) \cdot (10.0\mathbf{j} \text{ m}) = 7000 \text{ J}$$

His power output is
$$\mathcal{P} = \frac{W}{t} = \frac{7000 \text{ J}}{8.00 \text{ s}} = 875 \text{ W} \qquad \lozenge$$

---

41. A payload is fired from the Earth's surface with an initial speed of 2.00×10^4 m/s. What will its speed be when it is very far from the Earth? (Ignore friction.)

Solution   We apply the work-kinetic energy theorem between an initial point just after the payload is fired off at the Earth's surface, and a final point when it is coasting along far away.

$$K_i + W_{\text{other forces}} - f_k \Delta r = K_f$$

Section 6.9 of the text derives an expression for the work done by the varying gravitational force on the payload. The expression applies here with the mass of the Earth replacing the mass of the Sun:

$$\tfrac{1}{2} m_{\text{probe}} v_i^2 + G M_{\text{Earth}} m_{\text{probe}} \left(\frac{1}{r_f} - \frac{1}{r_i} \right) = \tfrac{1}{2} m_{\text{probe}} v_f^2$$

The reciprocal of the final distance is negligible compared with the reciprocal of the original distance from the Earth's center.

$$\tfrac{1}{2} v_i^2 + G M_{\text{Earth}} \left(\frac{1}{\infty} - \frac{1}{R_{\text{Earth}}} \right) = \tfrac{1}{2} v_f^2$$

We solve for the final speed, and substitute values tabulated on the endpapers:

$$v_f^2 = v_i^2 - \frac{2 G M_{\text{Earth}}}{R_{\text{Earth}}} = \left(2.00 \times 10^4 \text{ m/s} \right)^2 - \frac{2\left(6.67 \times 10^{-11} \text{ N} \cdot \text{m}^2/\text{kg}^2 \right)\left(5.98 \times 10^{24} \text{ kg} \right)}{6.37 \times 10^6 \text{ m}}$$

$$v_f^2 = 2.75 \times 10^8 \text{ m}^2/\text{s}^2 \qquad \text{and} \qquad v_f = 1.66 \times 10^4 \text{ m/s} \qquad \lozenge$$

---

45. A 4.00-kg particle moves along the x axis. Its position varies with time according to $x = t + 2.0t^3$, where x is in meters and t is in seconds. Find (a) the kinetic energy at any time t, (b) the acceleration of the particle and the force acting on it at time t, (c) the power being delivered to the particle at time t, and (d) the work done on the particle in the interval $t = 0$ to $t = 2.00$ s.

Solution Given $m = 4.00$ kg and $x = t + 2.0t^3$, we find

(a) $v = \dfrac{dx}{dt} = \dfrac{d}{dt}\left(t + 2.0t^3\right) = 1.00 + 6.0t^2$

$K = \dfrac{1}{2}mv^2 = \dfrac{1}{2}(4.00 \text{ kg})\left(1.00 + 6.0t^2\right)^2 = \left(2.00 + 24t^2 + 72t^4\right)$ J ◊

(b) $a = \dfrac{dv}{dt} = \dfrac{d}{dt}\left(1.00 + 6.0t^2\right) = 12t \text{ m / s}^2$ ◊

$F = ma = 4.00(12t) = 48t \text{ N}$ ◊

(c) $\mathcal{P} = \dfrac{dW}{dt} = \dfrac{dK}{dt} = \dfrac{d}{dt}\left(2.00 + 24t^2 + 72t^4\right) = \left(48t + 288t^3\right) \text{ W}$ ◊

[or use $\mathcal{P} = Fv - 48t\left(1.00 + 6.0t^2\right)$]

(d) $W = K_f - K_i$ where $t_i = 0$ and $t_f = 2.00$ s

At $t_i = 0$, $K_i = 2.00$ J At $t_f = 2.00$ s, $K_f = [2.00 + 24\,(2.00)^2 + 72\,(2.00)^4] = 1250$ J

Therefore, $W = 1248 \text{ J} = 1.25 \times 10^3 \text{ J}$ ◊

[or use $W = \displaystyle\int_{t_i}^{t_f} \mathcal{P}\, dt = \int_0^2 \left(48t + 288t^3\right) dt$, etc.]

55. The ball launcher in a pinball machine has a spring that has a force constant of 1.20 N/cm (Fig. P6.55). The surface on which the ball moves is inclined 10.0° with respect to the horizontal. If the spring is initially compressed 5.00 cm, find the launching speed of a 100-g ball when the plunger is released. Friction and the mass of the plunger are negligible.

Figure P6.55

Solution Use the work-kinetic energy theorem.

$W_{net} = \Delta K$: $\quad W_s + W_g = \Delta K$: $\quad \frac{1}{2}kx^2 - mg\Delta r \sin 10.0° = \frac{1}{2}m(v_f^2 - v_i^2)$

Since $v_i = 0$, $\qquad v_f = \sqrt{\frac{2}{m}\left[\frac{1}{2}kx^2 - mg\Delta r \sin 10.0°\right]}$

In this case, $\qquad x = \Delta r = 5.00 \text{ cm} = 5.00 \times 10^{-2} \text{ m}$

and $\qquad k = 1.20 \text{ N/cm} = 120 \text{ N/m}$

$$v_f = \sqrt{\frac{2\left[\frac{1}{2}(120 \text{ N/m})(5.00 \times 10^{-2} \text{ m})^2 - (0.100 \text{ kg})(9.80 \text{ m/s}^2)(5.00 \times 10^{-2} \text{ m})(\sin 10.0°)\right]}{0.100 \text{ kg}}}$$

and $v_f = 1.68 \text{ m/s}$ ◊

Chapter 7

Potential Energy

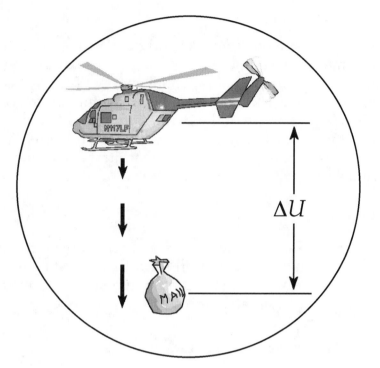

INTRODUCTION

In Chapter 6 we introduced the concept of kinetic energy, which is the energy associated with the motion of an object. In this chapter we introduce another form of mechanical energy, **potential energy**, which is the energy associated with the positions or configuration of interacting objects in a system. Potential energy can be thought of as stored energy that can either do work or be converted to kinetic energy.

The potential energy concept can be used only when dealing with a special class of forces called **conservative forces**. When only internal conservative forces, such as gravitational or spring forces, act within a system, the kinetic energy gained (or lost) by the system as its members change their relative positions is compensated by an equal energy loss (or gain) in potential energy. This is known as the **principle of conservation of mechanical energy**.

NOTES FROM SELECTED CHAPTER SECTIONS

7.1   Potential Energy of a System

Consider a system consisting of an object and the Earth. If an external agent raises the object through some distance, the work done by the external agent on the system is equal to the change in gravitational potential energy of the system. The gravitational potential energy associated with an object depends only on the object's weight and its vertical height above the surface of the Earth. If the height above the surface increases, the potential energy will also increase. In working problems involving gravitational potential energy, it is necessary to choose an arbitrary reference level (or location) at which the potential energy is zero.

7.2   The Isolated System

In an isolated system, the total mechanical energy (sum of the kinetic and potential energies) remains constant. In equation form this is a statement of conservation of mechanical energy, $\Delta K + \Delta U = 0$. This condition holds when the only forces doing the work on the system are conservative forces.

7.3   Conservative and Nonconservative Forces

A force is said to be **conservative** if the work done by that force on a body moving between any two points is independent of the path taken. In addition, the work done by a conservative force is zero when the body moves through any closed path and returns to its initial position. **Nonconservative** forces are those for which the work done on a particle moving between two points depends on the path. Furthermore, the work done by a nonconservative force (e.g. friction) around a closed path is not zero.

Conservative forces among members of a system cause no transformation of mechanical energy to internal energy within the system. The work done by a conservative force does not depend upon the path followed by members of the system; all the work depends only on the initial and final configurations of the system.

7.4   Conservative Forces and Potential Energy

It is possible to define a **potential energy function** associated with a conservative force. In a system of objects interacting via a conservative force, the work done by the conservative force on an object of the system that undergoes a displacement equals the negative of the change in the potential energy of the system associated with the force.

7.5   The Nonisolated System in Steady State

In an **isolated** system, no energy transfer occurs across the system boundary; the energy associated with a **nonisolated** system changes due to energy transfers across the boundary. When the nonisolated system is in steady state, the rates of energy and leaving the system are equal; the total energy of the system remains constant.

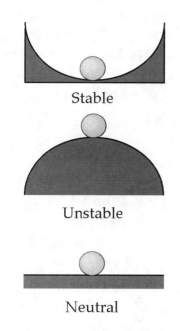

Stable

Unstable

Neutral

7.7   Energy Diagrams and Stability of Equilibrium

Positions of stable equilibrium correspond to points for which $U(x)$ has a relative minimum value on an energy diagram. Positions of unstable equilibrium correspond to those points for which $U(x)$ has a relative maximum value on an energy diagram. Finally, a position of neutral equilibrium corresponds to a region over which $U(x)$ remains constant.

Figure 7.1 Examples of stable, unstable, and neutral equilibrium.

EQUATIONS AND CONCEPTS

The gravitational potential energy associated with an object at any point in a gravitational field is the product of the object's weight and the vertical coordinate.

$$U_g \equiv mgy \qquad (7.2)$$

In a system of gravitationally interacting objects, the work done by the gravitational force as one object undergoes a displacement is equal to the negative of the change in the gravitational potential energy of the system.

$$W_g = U_i - U_f$$

The units of energy (kinetic and potential) are the same as the units of work — joules.

In calculating the work done by the gravitational force, remember that the difference in potential energy between two points is independent of the location of the origin. Choose an origin which is convenient to calculate U_i and U_f for a particular situation.

The law of **conservation of mechanical energy** says that if only conservative forces act on a system, the sum of the kinetic and potential energies of the system remains constant, or is conserved. According to this important conservation law, if the kinetic energy of the system increases by some amount, the potential energy must decrease by the same amount--and vice versa.

$$K_i + U_i = K_f + U_f \qquad (7.6)$$

If more than one conservative force acts on an object, then a potential energy term is associated with each force.

$$K_i + \Sigma U_i = K_f + \Sigma U_f$$

If the only conservative force acting within a system is the gravitational force, the equation for conservation of mechanical energy takes a special form.

$$\tfrac{1}{2}mv_i^2 + mgy_i = \tfrac{1}{2}mv_f^2 + mgy_f$$

Conservation of mechanical energy for a mass-spring system is similar:

$$\tfrac{1}{2}mv_i^2 + \tfrac{1}{2}kx_i^2 = \tfrac{1}{2}mv_f^2 + \tfrac{1}{2}kx_f^2$$

Total mechanical energy of a system is the sum of the kinetic and potential energies.

$$E_{mech} \equiv K + U \qquad (7.7)$$

Note that both the gravitational force and the spring force satisfy the required properties of a conservative force. That is, the work done is path independent and is zero for any closed path.

The quantity $\frac{1}{2}kx^2$ is referred to as the **elastic potential energy** stored in a spring which has been deformed a distance x from the equilibrium position.

$$U_s \equiv \tfrac{1}{2}kx^2 \qquad (7.8)$$

When nonconservative forces (such as friction) act within a system, the result is a change in the mechanical energy of the system.

$$K_i + U_i - f_k\Delta x = K_f + U_f \qquad (7.9)$$

A particular form of the continuity equation for energy applies to a system of several objects and includes any work done by an external force to change the energy of the system.

$$\Sigma K_i + \Sigma U_i - f_k\Delta x + W_{external} = \Sigma K_f + \Sigma U_f$$

A potential energy function U can be defined for a conservative force.

$$U_f = -\int_{x_i}^{x_f} F_x\,dx + U_i \qquad (7.14)$$

There is an important relationship between a conservative force and the potential energy of the system within which the force acts. The conservative force equals the negative derivative of the system's potential energy with respect to x.

$$F_x = -\frac{dU}{dx} \qquad (7.15)$$

SUGGESTIONS, SKILLS, AND STRATEGIES

Choosing a Zero Level

In working problems involving gravitational potential energy, it is always necessary to choose a system configuration in which the gravitational potential energy is zero. This choice is completely arbitrary because the important quantity is the **difference** in potential energy, and that difference is independent of the location of zero. It is often convenient, but not essential, to choose the reference configuration for zero potential energy with the moving object at the surface of the Earth. In most cases, the statement of the problem suggests a convenient configuration to use.

Conservation of Energy

Take the following steps in applying the principle of conservation of energy:

- Define your system, which may consist of more than one object. Determine if any energy transfers occur across the boundary. If so, use $\Delta E_{system} = \Sigma H$. If not, use $\Delta E_{system} = 0$.

- Select a reference position for the zero point of each form of potential energy.

- Determine whether or not nonconservative forces (e.g. friction or air resistance) are present.

- If mechanical energy of the system is conserved (that is, if only conservative forces are to act within the system), you can write the total initial energy at some point as the sum of the kinetic and potential energies at that point. Then, write an expression for the total final energy, $K_f + U_f$, at the final point of interest. Since mechanical energy is conserved, you can equate the two total energies and solve for the unknown.

- If nonconservative forces such as friction act within the system (and thus mechanical energy of the system is not conserved), first write expressions for the total initial and total final energies. In this case, the difference between the two total energies is equal to the mechanical energy gained or lost due to nonconservative force(s).

REVIEW CHECKLIST

▷ Recognize that the gravitational potential energy function, $U_g = mgy$, can be positive, negative, or zero, depending on the location of the reference level used to measure y. Be aware of the fact that although U depends on the origin of the coordinate system, the **change** in potential energy, $U_f - U_i$, is **independent** of the coordinate system used to define U.

▷ Understand that a force is said to be **conservative** if the work done by that force on a body moving between any two points is independent of the path taken. **Nonconservative** forces are those for which the work done on a particle moving between two points depends on the path. Account for nonconservative forces acting on a system by using the continuity equation for energy. In this case, the work done by all nonconservative forces equals the change in total mechanical energy of the system.

▷ Understand the distinction between kinetic energy (energy associated with motion), potential energy (energy associated with the positions or configuration of objects in a system), and the total mechanical energy of a system. State the law of conservation of mechanical energy, noting that mechanical energy is conserved when only conservative forces act within a system. This extremely powerful concept is important in all areas of physics.

ANSWERS TO SELECTED CONCEPTUAL QUESTIONS

1.   One person drops a ball from the top of a building while another person at the bottom observes its motion. Will these two people agree on the value of the gravitational potential energy of the ball-Earth system? On the change in potential energy? On the kinetic energy?

Answer   The two will not necessarily agree on the potential energy, since this depends on the origin--which may be chosen differently for the two observers. However, the two **must** agree on the value of the **change** in potential energy, which is independent of the choice of the reference frames. The two will also agree on the kinetic energy of the ball, assuming both observers are at rest with respect to each other, and hence measure the same v.

□   □   □   □

9. You ride a bicycle. In what sense is your bicycle solar-powered?

Answer The energy to ride the bicycle comes from your body. The source of that energy is the food that you ate at some previous time. The energy in the food, assuming that we focus on vegetables, came from the growth of the plant, for which photosynthesis is a major factor. The light for the photosynthesis comes from the Sun. The argument for meats has a couple of extra steps, but also goes through the process of photosynthesis in the plants eaten by animals. Thus, the source of the energy to ride the bicycle is the Sun, and your bicycle is solar-powered!

□   □   □   □

10. A bowling ball is suspended from the ceiling of a lecture hall by a strong cord. The bowling ball is drawn away from its equilibrium position and released from rest at the tip of the demonstrator's nose. If the demonstrator remains stationary, explain why she will not be struck by the ball on its return swing. Would the demonstrator be safe if she pushed the ball as she released it?

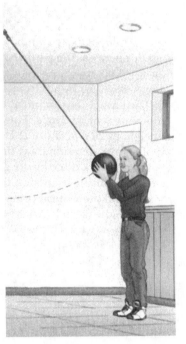

Answer The total energy of the system (the bowling ball and the Earth) must be conserved. Since the system initially has a potential energy mgh, and the ball has no kinetic energy, it cannot have any kinetic energy when returning to its initial position. Of course, air resistance will cause the ball to return to a point slightly below its initial position. On the other hand, if the ball is given a push, the demonstrator's nose will be in big trouble.

□   □   □   □

SOLUTIONS TO SELECTED END-OF-CHAPTER PROBLEMS

5. A bead slides without friction around a loop-the-loop (Fig. P7.5). The bead is released from a height $h = 3.50R$. What is its speed at point A? How large is the normal force on it at point A if its mass is 5.00 g?

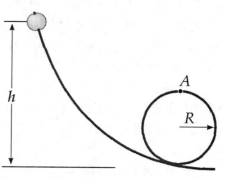

Solution Choose the system as bead, wire, and Earth. Use the energy version of the isolated system model. It is convenient to choose the reference configuration for potential energy to be with the bead at the lowest point of its motion.

Figure P7.5

Since $v_i = 0$, $\qquad E_i = K_i + U_i = 0 + mgh = mg(3.50R)$

The total energy of the bead-Earth system when the bead is at point A can be written as

$$E_A = K_A + U_A = \tfrac{1}{2}mv_A^2 + mg(2R)$$

Since mechanical energy is conserved, $E_i = E_A$, and we get

$$\tfrac{1}{2}mv_A^2 + mg(2R) = mg(3.50R)$$

$$v_A^2 = 3.00gR \qquad \text{or} \qquad v_A = \sqrt{3.00gR} \qquad \qquad \Diamond$$

To find the normal force at the top, it is useful to construct a free-body diagram as shown, where both **n** and mg are downward. Newton's second law gives

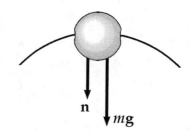

$$n + mg = \frac{mv_A^2}{R} = \frac{m(3.00gR)}{R} = 3.00mg$$

$$n = 3.00mg - mg = 2.00mg$$

$$n = 2.00(5.00 \times 10^{-3} \text{ kg})(9.80 \text{ m} / \text{s}^2) = 0.0980 \text{ N} \quad \text{downward} \qquad \qquad \Diamond$$

9. Two objects are connected by a light string passing over a light, frictionless pulley as shown in Figure P7.9. The 5.00-kg object is released from rest. Using the law of conservation of energy, (a) determine the speed of the 3.00-kg object just as the 5.00-kg object hits the ground. (b) Find the maximum height to which the 3.00-kg object rises.

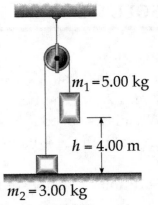

$m_1 = 5.00$ kg

$h = 4.00$ m

Solution   As the system choose the two blocks A and B, string, pulley, and Earth.

$m_2 = 3.00$ kg

(a)   Choose the initial point before release and the final point just before the larger object hits the floor. The total energy of the system remains constant and the energy version of the isolated system model gives

Figure P7.9

$$(K_A + K_B + U_g)_i = (K_A + K_B + U_g)_f$$

At the initial point K_{Ai} and K_{Bi} are zero and we define the gravitational potential energy of the system as zero. No external forces do work on the system and no friction acts within the system. Thus,

$$0 = \tfrac{1}{2}(5.00 \text{ kg} + 3.00 \text{ kg})v_f{}^2 + (3.00 \text{ kg})(9.80 \text{ m/s}^2)(4.00 \text{ m}) + (5.00 \text{ kg})(9.80 \text{ m/s}^2)(-4.00 \text{ m})$$

$$(2.00 \text{ kg})(9.80 \text{ m/s}^2)(4.00 \text{ m}) = \tfrac{1}{2}(8.00 \text{ kg})v_f{}^2$$

$$v_f = 4.43 \text{ m/s} \qquad \lozenge$$

(b)   Now the string goes slack. The 3.00-kg object becomes a projectile. We focus now on the system of the 3.00-kg object and the Earth. Take the initial point at the previous final point, and the new final point at its maximum height: $(K + U_g)_i = (K + U_g)_f$

$$\tfrac{1}{2}(3.00 \text{ kg})(4.43 \text{ m/s})^2 + (3.00 \text{ kg})(9.80 \text{ m/s}^2)(4.00 \text{ m}) = 0 + (3.00 \text{ kg})(9.80 \text{ m/s}^2)y_f$$

$$y_f = 5.00 \text{ m} \qquad \lozenge$$

or 1.00 m higher than the height of the 5.00-kg mass when it was released.

13. A force acting on a particle moving in the xy plane is given by $\mathbf{F} = (2y\mathbf{i} + x^2\mathbf{j})$ N, where x and y are in meters. The particle moves from the origin to a final position having coordinates $x = 5.00$ m and $y = 5.00$ m, as in Figure P7.12. Calculate the work done by $\mathbf{F}$ along (a) OAC, (b) OBC, (c) OC. (d) Is $\mathbf{F}$ conservative or nonconservative? Explain.

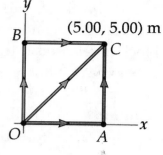

Figure P7.12

Solution

In the following integrals, remember $\mathbf{i} \cdot \mathbf{i} = \mathbf{j} \cdot \mathbf{j} = 1$ and $\mathbf{i} \cdot \mathbf{j} = 0$.

(a) $W_{OA} = \int_0^{5.00} (2y\mathbf{i} + x^2\mathbf{j}) \cdot (\mathbf{i}\,dx) = \int_0^{5.00} 2y\,dx = 2y\int_0^{5.00} dx = 2yx \Big]_{x=0,y=0}^{x=5.00,y=0} = 0$

$W_{AC} = \int_0^{5.00} (2y\mathbf{i} + x^2\mathbf{j}) \cdot (\mathbf{j}\,dy) = \int_0^{5.00} x^2\,dy = x^2 \int_0^{5.00} dy = x^2 y \Big]_{x=5.00,y=0}^{x=5.00,y=5.00} = 125 \text{ J}$

$W_{OAC} = 0 + 125 \text{ J} = 125 \text{ J}$ ◊

(b) $W_{OB} = \int_0^{5.00} (2y\mathbf{i} + x^2\mathbf{j}) \cdot (\mathbf{j}\,dy) = \int_0^{5.00} x^2\,dy = x^2 \int_0^{5.00} dy = x^2 y \Big]_{x=0...}^{x=0...} = 0$

$W_{BC} = \int_0^{5.00} (2y\mathbf{i} + x^2\mathbf{j}) \cdot (\mathbf{i}\,dx) = \int_0^{5.00} 2y\,dx = 2y\int_0^{5.00} dx = 2(5.00)x \Big]_0^{5.00} = 50.0 \text{ J}$

$W_{OBC} = 0 + 50.0 \text{ J} = 50.0 \text{ J}$ ◊

(c) $W_{OC} = \int (2y\mathbf{i} + x^2\mathbf{j}) \cdot (\mathbf{i}\,dx + \mathbf{j}\,dy) = \int (2y\,dx + x^2\,dy)$

Since $x = y$ along OC, $dx = dy$ and

$W_{OC} = \int_0^{5.00} (2x + x^2)dx = 66.7 \text{ J}$ ◊

(d) $\mathbf{F}$ is non-conservative since the work done is path dependent. ◊

19. A block of mass 0.250 kg is placed on top of a light vertical spring of constant $k = 5000$ N/m and is pushed downward so that the spring is compressed 0.100 m. After the block is released it travels upward and then leaves the spring. To what maximum height above the point of release does it rise?

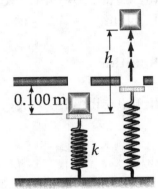

Solution In both the initial and final states, the block is not moving. Therefore, the initial and final energies of the block-spring-Earth system are:

$$E_i = K_i + U_i = 0 + (U_g + U_s)_i = 0 + \left(0 + \frac{1}{2}kx^2\right)$$

$$E_f = K_f + U_f = 0 + (U_g + U_s)_f = 0 + (mgh + 0)$$

Since $E_i = E_f$, $\qquad mgh = \frac{1}{2}kx^2$

and $\qquad h = \dfrac{kx^2}{2mg} = \dfrac{(5000 \text{ N}/\text{m})(0.100 \text{ m})^2}{2(0.250 \text{ kg})(9.80 \text{ m}/\text{s}^2)} = 10.2 \text{ m}$ $\qquad\qquad\Diamond$

---

21. The coefficient of friction between the 3.00-kg block and surface in Figure P7.21 is 0.400. The system starts from rest. What is the speed of the 5.00-kg ball when it has fallen 1.50 m?

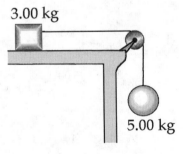

Figure P7.21

Solution We could solve this problem by using $\Sigma F = ma$ to give a pair of simultaneous equations in the unknown acceleration and tension; then we would have to solve a motion problem to find the final speed. It is easier to solve using the energy version of the isolated system model.

For the 3.00-kg block, $\qquad F_g = mg = (3.00 \text{ kg})(9.80 \text{ m}/\text{s}^2) = 29.4 \text{ N}$

From $\quad \Sigma F_y = ma_y$, $\qquad n - 29.4 \text{ N} = 0$

and $\quad n = 29.4 \text{ N}$ $\qquad f_k = \mu_k n = 0.400(29.4 \text{ N}) = 11.8 \text{ N}$

Now for the block-ball-Earth system, the continuity equation for energy is

$$(K_A + K_B + U_g)_i + W_{external} - f_k \Delta x = (K_A + K_B + U_g)_f \qquad\qquad (1)$$

Choose the initial point before release and the final point after each object has moved 1.50 m. Choose the zero of potential energy for the system when the ball is located 1.50 m below its initial position and the block is on the tabletop.

By substituting into Eq. (1):

$$0+0+0+m_B g y_{Bi} - f\Delta x = \tfrac{1}{2}m_A v_f^2 + \tfrac{1}{2}m_B v_f^2 + 0 + 0$$

$$(5.00\text{ kg})(9.80\text{ m/s}^2)(1.50\text{ m}) - (11.8\text{ N})(1.50\text{ m}) = \tfrac{1}{2}(3.00\text{ kg})v_f^2 + \tfrac{1}{2}(5.00\text{ kg})v_f^2$$

$$73.5\text{ J} - 17.6\text{ J} = \tfrac{1}{2}(8.00\text{ kg})v_f^2 \quad \text{or} \quad v_f = \sqrt{\frac{2(55.9\text{ J})}{8.00\text{ kg}}} = 3.74\text{ m/s} \qquad \Diamond$$

---

23. A 5.00-kg block is set into motion up an inclined plane with an initial speed of 8.00 m/s (Fig. P7.23). The block comes to rest after traveling 3.00 m along the plane, which is inclined at an angle of 30.0° to the horizontal. For this motion, determine (a) the change in the block's kinetic energy, (b) the change in potential energy of the block-Earth system, and (c) the friction force exerted on the block (assumed to be constant). (d) What is the coefficient of kinetic friction?

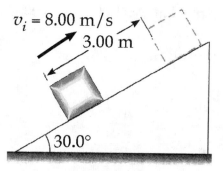

Figure P7.23

Solution

(a) $\Delta K = K_f - K_i = \tfrac{1}{2}mv_f^2 - \tfrac{1}{2}mv_i^2 = 0 - \tfrac{1}{2}(5.00\text{ kg})(8.00\text{ m/s})^2 = -160\text{ J}$ $\qquad \Diamond$

(b) $\Delta U = U_{gf} - U_{gi} = mgy_f - 0 = (5.00\text{ kg})(9.80\text{ m/s}^2)(3.00\text{ m})\sin 30.0° = 73.5\text{ J}$ $\qquad \Diamond$

(c) $(K+U_g)_i + W_{external} - f_k\Delta x = (K+U_g)_f$

$$\tfrac{1}{2}(5.00\text{ kg})(8.00\text{ m/s})^2 + 0 - f(3.00\text{ m}) = 0 + (5.00\text{ kg})(9.80\text{ m/s}^2)(1.50\text{ m})$$

$$160\text{ J} - f(3.00\text{ m}) = 73.5\text{ J} \qquad \text{and} \qquad f = \frac{86.5\text{ J}}{3.00\text{ m}} = 28.8\text{ N} \qquad \Diamond$$

(d) The forces perpendicular to the incline must add to zero.

$\Sigma F_y = 0:$ $+n - mg \cos 30.0° = 0$

Substituting, $n = (5.00 \text{ kg})(9.80 \text{ m}/\text{s}^2) \cos 30.0° = 42.4 \text{ N}$

Now, $f_k = \mu_k n$ gives $\mu_k = \dfrac{f_k}{n} = \dfrac{28.8 \text{ N}}{42.4 \text{ N}} = 0.679$ ◊

---

27. A single conservative force acts on a 5.00-kg particle. The force equation $F_x = (2x + 4) \text{ N}$ describes the force, where x is in meters. As the particle moves along the x-axis from $x = 1.00 \text{ m}$ to $x = 5.00 \text{ m}$, calculate (a) the work done by this force, (b) the change in the potential energy of the system, and (c) the kinetic energy of the particle at $x = 5.00 \text{ m}$ if its speed at $x = 1.00 \text{ m}$ is 3.00 m/s.

Solution

(a) $W_F = \displaystyle\int_{x_i}^{x_f} F_x \, dx$

where $F_x = (2x + 4) \text{ N}, \quad x_i = 1.00 \text{ m}, \quad \text{and} \quad x_f = 5.00 \text{ m}$

therefore, $W_F = \displaystyle\int_{1.00 \text{ m}}^{5.00 \text{ m}} (2x + 4) dx \text{ N} \cdot \text{m} = x^2 + 4x \Big]_{1.00 \text{ m}}^{5.00 \text{ m}} \text{ N} \cdot \text{m} = 40.0 \text{ J}$ ◊

(b) The change in potential energy equals the negative of the work done by the conservative force.

$$\Delta U = -W_F = -40.0 \text{ J}$$ ◊

(c) When only conservative forces act, conservation of energy gives $K_i + U_i = K_f + U_f$.

Rearranging, $K_f = K_i - (U_f - U_i) = \tfrac{1}{2} m v_i^2 - \Delta U$

so $K_f = \left(\tfrac{1}{2}\right)(5.00 \text{ kg})(3.00 \text{ m}/\text{s})^2 - (-40.0 \text{ J}) = 62.5 \text{ J}$ ◊

---

29. The potential energy of a system of two particles separated by a distance r is given by $U(r) = A/r$, where A is a constant. Find the radial force $\mathbf{F}_r$ that each particle exerts on the other.

Solution The force is the negative derivative of the potential energy with respect to distance:

$$F_r = -\frac{dU}{dr} = -\frac{d}{dr}\left(Ar^{-1}\right) = -A(-1)r^{-2} = \frac{A}{r^2} \qquad \Diamond$$

This describes an inverse-square-law force of repulsion, as between two negative point electric charges.

---

43. The particle described in Problem 42 (Fig. P7.42) is released from rest at A and the surface of the bowl is rough. The speed of the particle at B is 1.50 m/s. (a) What is its kinetic energy at B? (b) How much mechanical energy is transformed due to friction as the particle moves from A to B? (c) Is it possible to determine the coefficient of friction from these results in any simple manner? Explain.

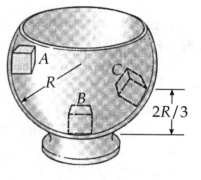

Solution Let us take $U = 0$ for the particle-bowl-Earth system when the particle is at B.

Figure P7.42

Since $v_i = 0$ at A, $\qquad K_A = 0 \qquad$ and $\qquad U_A = mgR$

(a) Since $\qquad\qquad v_B = 1.50$ m/s $\qquad$ and $\qquad m = 200$ g

$$K_B = \tfrac{1}{2}mv_B^2 = \tfrac{1}{2}(0.200 \text{ kg})(1.50 \text{ m/s})^2 = 0.225 \text{ J} \qquad \Diamond$$

(b) At A, $\qquad E_i = K_A + U_A = 0 + mgR = (0.200 \text{ kg})(9.80 \text{ m/s}^2)(0.300 \text{ m}) = 0.588 \text{ J}$

At B, $\qquad E_f = K_B + U_B = 0.225 \text{ J}$

The decrease in mechanical energy is equal to the increase in internal energy.

$$E_{mech,\, i} - \Delta E_{int} = E_{mech,\, f}$$

The energy transformed is $\qquad \Delta E_{int} = -\Delta E_{mech} = E_i - E_f = 0.588 \text{ J} - 0.225 \text{ J} = 0.363 \text{ J} \Diamond$

(c) Even though the energy transformed is known, both the normal force and the friction force change with position as the block slides on the inside of the bowl. Therefore, there is no easy way to find the coefficient of friction. $\qquad \Diamond$

---

47. A 10.0-kg block is released from point A in Figure P7.47. The track is frictionless except for the portion between points B and C, which has a length of 6.00 m. The block travels down the track, hits a spring of force constant $k = 2250$ N/m, and compresses the spring 0.300 m from its equilibrium position before coming to rest momentarily. Determine the coefficient of kinetic friction between the block and the rough surface between B and C.

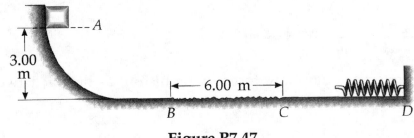

Figure P7.47

Solution Choose the initial point when the block is at A and the final point when the spring is fully compressed.

$$(K + U_g + U_s)_A - f_k \Delta x = (K + U_g + U_s)_f$$

$$0 + mgy_A + 0 - f_k \Delta x_{BC} = 0 + 0 + \tfrac{1}{2}kx_f^2$$

$$(10.0 \text{ kg})(9.80 \text{ m/s}^2)(3.00 \text{ m}) - f(6.00 \text{ m}) = \tfrac{1}{2}(2250 \text{ N/m})(0.300 \text{ m})^2$$

$$294 \text{ J} - f(6.00 \text{ m}) = 101 \text{ J} \qquad \text{so} \qquad f = 193 \text{ J}/6.00 \text{ m} = 32.1 \text{ N}$$

Now consider the vertical forces when the block is between B and C.

From $\Sigma F_y = 0$, $\qquad\qquad +n - mg = 0 \qquad$ and $\qquad n = (10.0 \text{ kg})(9.80 \text{ m/s}^2) = 98.0 \text{ N}$

Applying $f_k = \mu_k n$: $\qquad \mu_k = f_k/n = 32.1 \text{ N}/98.0 \text{ N} = 0.328$ ◊

Another way to look at the problem is to say that the original potential energy mgh is transformed into an increase in internal energy, and into compressing the spring:

$$mgh = f\Delta x + \tfrac{1}{2}kx^2 \qquad \text{or} \qquad f\Delta x = \mu mgd = mgh - \tfrac{1}{2}kx^2$$

Substituting $m = 10.0$ kg, $h = 3.00$ m, $k = 2250$ N/m, $x = 0.300$ m, $d = \Delta x = 6.00$ m

yields $\qquad\qquad \mu = \dfrac{h}{d} - \dfrac{kx^2}{2mgd} = 0.328$ ◊

49. A 20.0-kg block is connected to a 30.0-kg block by a string that passes over a frictionless pulley. The 30.0-kg block is connected to a spring that has negligible mass and a force constant of 250 N/m, as in Figure P7.49. The spring is unstretched when the system is as shown in the figure, and the incline is frictionless. The 20.0-kg block is pulled 20.0 cm down the incline (so that the 30.0-kg block is 40.0 cm above the floor) and released from rest. Find the speed of each block when the 30.0-kg block is again 20.0 cm above the floor (i.e., when the spring is unstretched).

Solution   Let x be the distance the spring is stretched from equilibrium ($x = 0.200$ m), which corresponds to the upward displacement of the 30.0-kg mass. Also let $U_g = 0$ be measured with respect to the lowest position of the 20.0-kg mass when the system is released from rest. Finally, define v as the speed of both blocks at the moment the spring passes through its unstretched position. Since all forces are conservative, conservation of energy yields

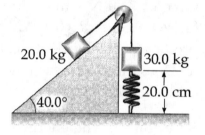

Figure P7.49

$$\Delta K + \Delta U_s + \Delta U_g = 0$$

Solving for each variable,

$$\Delta K = \frac{1}{2}(m_1 + m_2)v^2 - 0 = \frac{1}{2}(50.0 \text{ kg})v^2 = (25.0 \text{ kg})v^2$$

$$\Delta U_s = 0 - \frac{1}{2}kx^2 = -\frac{1}{2}(250 \text{ N/m})(0.200 \text{ m})^2 = -5.00 \text{ N} \cdot \text{m}$$

$$\Delta U_g = (m_2 \sin\theta - m_1)gx = [(20.0 \text{ kg})\sin 40.0° - 30.0\text{kg}](9.80 \text{ m/s}^2)(0.200 \text{ m}) = -33.6 \text{ N} \cdot \text{m}$$

Substituting into our equation representing the energy version of the isolated system model,

$$(25.0 \text{ kg})v^2 - 5.00 \text{ N} \cdot \text{m} - 33.6 \text{ N} \cdot \text{m} = 0$$

Solving for v gives $\qquad\qquad\qquad\qquad v = 1.24 \text{ m/s}$ ◊

51. A block of mass 0.500 kg is pushed against a horizontal spring of negligible mass until the spring is compressed a distance of Δx (Fig. P7.51). The force constant of the spring is 450 N/m. When released, the block travels along a frictionless, horizontal surface to point B, the bottom of a vertical circular track of radius $R = 1.00$ m, and continues to move up the track. The speed of the block at the bottom of the track is $v_B = 12.0$ m/s, and the block experiences an average frictional force of 7.00 N while sliding up the track. (a) What is Δx? (b) What speed do you predict for the block at the top of the track? (c) Does the block reach the top of the track, or does it fall off before reaching the top?

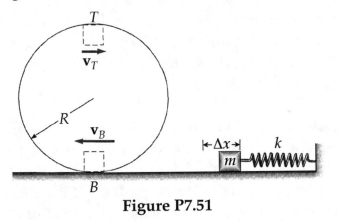

Figure P7.51

Solution   The energy of the block-spring system is conserved in the firing of the block. Therefore,

(a) $\frac{1}{2}kx^2 = \frac{1}{2}mv^2$ or $\frac{1}{2}(450 \text{ N / m})(\Delta x)^2 = \frac{1}{2}(0.500 \text{ kg})(12.0 \text{ m / s})^2$

Thus, $\Delta x = 0.400$ m   ◊

(b)   To find speed of block at the top, we consider the block-Earth system.

$(K + U_g)_B - f_k \Delta x = (K + U_g)_T$: $\left(mgh_B + \frac{1}{2}mv_B^2\right) - f(\pi R) = \left(mgh_T + \frac{1}{2}mv_T^2\right)$

Substituting, $mgh_T = (0.500 \text{ kg})(9.80 \text{ m / s}^2)(2.00 \text{ m}) = 9.80$ J

We have $0 + \frac{1}{2}(0.500 \text{ kg})(12.0 \text{ m / s})^2 - (7.00 \text{ N})(\pi)(1.00 \text{ m}) = 9.80 \text{ J} + \frac{1}{2}(0.500 \text{ kg})v_T^2$

and $0.250 v_T^2 = 4.21$

Thus, $v_T = 4.10$ m/s   ◊

(c)   The block falls if $a_c < g$ $a_c = \dfrac{v_T^2}{R} = \dfrac{(4.10 \text{ m / s})^2}{1.00 \text{ m}} = 16.8 \text{ m / s}^2$

Therefore, $a_c > g$. Some downward normal force is required along with the block's weight to provide the centripetal acceleration, and the block stays on the track.   ◊

Chapter 8

Momentum and Collisions

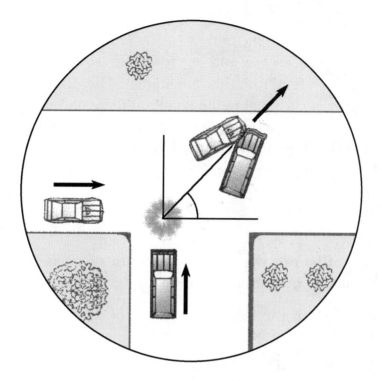

INTRODUCTION

One of the main objectives of this chapter is to enable you to understand and analyze collisions and other events in which objects experience large accelerations as a result of very large forces which act for a short time interval. As a first step, we shall introduce the concept of **momentum**, a term that is used in describing objects in motion. Momentum is defined as the product of mass and velocity.

The concept of momentum leads us to a second conservation law, that of conservation of momentum. This law is especially useful for treating problems that involve collisions between objects and for analyzing rocket propulsion. The concept of the center of mass of a system of particles is also introduced, and we shall see that the motion of a system of particles can be modeled as the motion of one representative particle located at the center of mass.

NOTES FROM SELECTED CHAPTER SECTIONS

8.2   Impulse and Momentum

If two particles of masses m_1 and m_2 form an **isolated system**, then the total momentum of the system remains constant.   The **time rate of change of the momentum** of a particle is equal to the resultant force on the particle. The **impulse** of a force equals the change in momentum of the particle on which the force acts. Under the **impulse approximation,** it is assumed that one of the forces acting on a particle is of short time duration but of much greater magnitude than any of the other forces.

8.3   Collisions

For **any type of collision,** the total momentum of an isolated system before the collision equals the total momentum just after the collision.

In an **inelastic collision,** the total momentum of an isolated system is conserved; however, the total kinetic energy is not conserved.

In a **perfectly inelastic collision,** the two colliding objects stick together following the collision.

In an **elastic collision,** both momentum and kinetic energy of an isolated system are conserved.

8.4   Two-Dimensional Collisions

The law of conservation of momentum is not restricted to one-dimensional collisions. If two masses undergo a **two-dimensional** (glancing) **collision** and there are no external forces acting, the total momentum in each of the x, y, and z directions is conserved.

8.5   The Center of Mass

The position of the center of mass of a system can be described as the average position of the system's mass. If **g** is constant over a mass distribution, then the center of gravity will coincide with the center of mass.

8.6 Motion of a System of Particles

The center of mass of a system of particles moves like an imaginary particle of mass M (equal to the total mass of the system) under the influence of the resultant external force on the system.

8.7 Rocket Propulsion

The operation of a rocket depends on the law of conservation of linear momentum applied to a system of particles (the rocket plus its ejected fuel). The thrust of the rocket is the force exerted on it by the exhaust gases.

EQUATIONS AND CONCEPTS

The **linear momentum p** of a particle is defined as the product of its mass m with its velocity **v**. This equation is equivalent to three component scalar equations, one along each of the coordinate axes.

$$\mathbf{p} = m\mathbf{v} \tag{8.1}$$

The time rate of change of the linear momentum of a particle is equal to the resultant force acting on the particle. This is Newton's second law for a particle.

$$\sum \mathbf{F} = \frac{d\mathbf{p}}{dt} \tag{8.3}$$

When two particles interact with each other (but are otherwise isolated from their surroundings), Newton's third law **tells us that the force of particle 1 on particle 2 is equal and opposite to the force of particle 2 on particle 1.** Since force is the time rate of change of momentum (Newton's second law), one finds that the total momentum of the isolated pair of particles is conserved.

$$\mathbf{p}_{tot} = \text{constant} \tag{8.4}$$

In general, when the **external force** acting on a system of particles is zero, the **total linear momentum of the system is conserved.** This important statement is known as the **law of conservation of momentum.** It is especially useful in treating problems involving collisions between two bodies.

$$\mathbf{p}_{1i} + \mathbf{p}_{2i} = \mathbf{p}_{1f} + \mathbf{p}_{2f} \qquad (8.5)$$

The **impulse** of a force $\mathbf{F}$ acting on a particle **equals the change in momentum of the particle.** This is known as the impulse-momentum theorem.

$$\mathbf{I} \equiv \int_{t_i}^{t_f} \sum \mathbf{F}\, dt = \Delta\mathbf{p} \qquad (8.9)$$

To calculate impulse, we usually define a **time-averaged force** $\overline{\mathbf{F}}$ which would give the same impulse to the particle as the actual time-varying force over the time interval Δt.

$$\overline{\mathbf{F}} \equiv \frac{1}{\Delta t}\int_{t_i}^{t_f} \sum \mathbf{F}\, dt \qquad (8.11)$$

The average force can be thought of as the constant force that would give the same change in momentum over the time interval Δt as the applied impulse.

$$\mathbf{I} = \Delta\mathbf{p} = \sum \overline{\mathbf{F}}\,\Delta t \qquad (8.12)$$

In the impulse approximation, the force $\mathbf{F}$ appearing in Equation 8.9 acts for a short time and is much larger than any other force present. This approximation is usually assumed in problems involving collisions, where the force is the contact force between the particles during the collision.

It is useful to consider two particular types of collisions that can occur between two bodies in an isolated system. **An elastic collision is one in which both linear momentum and kinetic energy are conserved.** An **inelastic collision is one in which only linear momentum is conserved.** A perfectly inelastic collision is an inelastic collision in which the two bodies stick together after the collision. Note that momentum of the system is conserved in any type of collision. Furthermore, note that when we say that the momentum is conserved, we are speaking about the momentum of the **entire system.** That is, the momentum of each particle may change as the result of the collision, but the momentum of the system remains unchanged.

$$\mathbf{p}_1 + \mathbf{p}_2 = \text{constant}$$
$$K_1 + K_2 = \text{constant}$$
(Elastic)

$$\mathbf{p}_1 + \mathbf{p}_2 = \text{constant} \qquad \text{(Inelastic)}$$

When two particles moving along a straight line collide and stick together (perfectly inelastic collision), the common velocity after the collision can be calculated in terms of the two mass values and the two initial velocities.

$$\mathbf{v}_f = \frac{m_1\mathbf{v}_{1i} + m_2\mathbf{v}_{2i}}{m_1 + m_2} \qquad (8.15)$$

When two particles undergo a one-dimensional perfectly elastic collision, both momentum and kinetic energy of the system are conserved. When such a collision occurs, **the relative velocity before the collision equals the negative of the relative velocity of the two particles following the collision.**

$$v_{1i} - v_{2i} = -\left(v_{1f} - v_{2f}\right) \qquad (8.20)$$

In an elastic collision between two objects, the final velocity components can be stated in terms of the initial velocities and masses. The signs of the velocity components are determined by the directions of motion.

$$v_{1f} = \left(\frac{m_1 - m_2}{m_1 + m_2}\right)v_{1i} + \left(\frac{2m_2}{m_1 + m_2}\right)v_{2i} \quad (8.21)$$

$$v_{2f} = \left(\frac{2m_1}{m_1 + m_2}\right)v_{1i} + \left(\frac{m_2 - m_1}{m_1 + m_2}\right)v_{2i} \quad (8.22)$$

The x-**coordinate of the center of mass of** n **particles** whose individual coordinates are x_1, x_2, x_3, . . . and whose masses are m_1, m_2, m_3, . . . is given by Equation 8.29. The y and z coordinates of the center of mass are defined by similar expressions. The center of mass of a homogeneous, symmetric body must lie on an axis of symmetry.

$$x_{CM} \equiv \frac{\sum_i m_i x_i}{\sum_i m_i} = \frac{\sum_i m_i x_i}{M} \quad (8.29)$$

The center of mass for a collection of particles can be located by its position vector.

$$\mathbf{r}_{CM} = \frac{1}{M}\sum_i m_i \mathbf{r}_i \quad (8.31)$$

In this expression for the **velocity of the center of mass of a system of particles**, v_i is the velocity of the ith particle and M is the total mass of the system.

$$\mathbf{v}_{CM} = \frac{1}{M}\sum_i m_i \mathbf{v}_i \quad (8.35)$$

The **acceleration of the center of mass of a system of particles** depends on the value of the acceleration for each of the individual particles.

$$\mathbf{a}_{CM} = \frac{1}{M}\sum_i m_i \mathbf{a}_i \quad (8.37)$$

Newton's second law applied to a system of particles says that the **resultant external force** acting on the system **equals the time rate of change of the total momentum.** The form of Newton's second law must be used when the mass of the system changes. If the net external force on the system is zero, then the total momentum of the system remains constant.

$$\Sigma \mathbf{F}_{\text{ext}} = M \mathbf{a}_{\text{CM}} = \frac{d\mathbf{p}_{\text{tot}}}{dt} \qquad (8.39)$$

The **total momentum of a system of particles** is equal to the total mass M multiplied by the velocity of the center of mass.

$$\mathbf{p}_{\text{tot}} = M\mathbf{v}_{\text{CM}} \qquad (8.40)$$

$$(\text{constant when } \Sigma \mathbf{F}_{\text{ext}} = 0)$$

The principle behind the operation of a rocket is the law of conservation of momentum as applied to the rocket and its ejected fuel. If a rocket moves in the absence of gravity and ejects fuel with an exhaust velocity v_e, **its change in velocity** is proportional to the exhaust velocity, where M_i and M_f refer to its initial and final mass values for the rocket.

$$v_f - v_i = v_e \ln\left(\frac{M_i}{M_f}\right) \qquad (8.42)$$

The thrust force on a rocket increases as the exhaust speed increases and as the burn rate increases.

$$\text{Instantaneous thrust} = \left| v_e \frac{dM}{dt} \right| \qquad (8.43)$$

SUGGESTIONS, SKILLS, AND STRATEGIES

The following procedure is recommended when dealing with problems involving collisions between two objects:

- Set up a coordinate system and define your velocities with respect to that system. That is, objects moving in the direction selected as the positive direction of the x axis are considered as having a positive velocity and negative if moving in the negative-x direction. It is convenient to have the x-axis coincide with one of the initial velocities.

- In your sketch of the coordinate system, draw all velocity vectors with labels and include all the given information.

- Write expressions for the momentum of each object before and after the collision. (In two-dimensional collision problems, write expressions for the x and y components of momentum before and after the collision.) Remember to include the appropriate signs for the velocity vectors.

- Now write expressions for the **total** momentum of the system of two objects **before** and **after** the collision and equate the two. (For two-dimensional collisions, this expression should be written for the momentum in both the x and y directions.) It is important to emphasize that it is the momentum of the **system** (the two colliding objects) that is conserved, not the momentum of the individual objects.

- If the collision is **inelastic**, you should then proceed to solve the momentum equations for the unknown quantities.

- If the collision is **elastic**, kinetic energy of the system is also conserved, so you can equate the total kinetic energy before the collision to the total kinetic energy after the collision. This gives an additional relationship between the various velocities. The conservation of kinetic energy for one-dimensional elastic collisions leads to the expression $v_{1i} - v_{2i} = -(v_{1f} - v_{2f})$, which is often easier to use in solving elastic collision problems than is an expression for conservation of kinetic energy.

REVIEW CHECKLIST

▷ The impulse of a force acting on a particle during some time interval equals the **change** in momentum of the particle, and the impulse equals the area under the force-time graph.

▷ The momentum of any isolated system (one for which the net external force is zero) is conserved, regardless of the nature of the forces between the masses which compose the system.

▷ There are two types of collisions that can occur between two particles, namely elastic and inelastic collisions. Recognize that a **perfectly** inelastic collision is an inelastic collision in which the colliding particles stick together after the collision, and hence move as a composite particle.

▷ The conservation of linear momentum applies not only to head-on collisions (one-dimensional), but also to glancing collisions (two- or three-dimensional). For example, in a two-dimensional collision, the total momentum in the x direction is conserved and the total momentum in the y direction is conserved.

▷ The equations for momentum and kinetic energy can be used to calculate the final velocities in a two-body head-on elastic collision; and to calculate the final velocity and the change of kinetic energy in a two-body system for a completely inelastic collision.

ANSWERS TO SELECTED CONCEPTUAL QUESTIONS

3. Explain how linear momentum is conserved when a ball bounces from a floor.

Answer   The ball's downward momentum increases as it accelerates downward. A larger upward momentum change occurs when it touches the floor and rebounds. The outside forces of gravity and the normal force inject impulses to change its momentum.

If we think of ball-and-Earth-together as our system, these forces are internal and do not change the total momentum. It is conserved, if we neglect the curvature of the Earth's orbit. As the ball falls down, the Earth lurches up to meet it, on the order of 10^{25} times more slowly. Then, ball and Earth bounce off each other and separate. In other words, while you dribble a ball, you also dribble the Earth.

10. A sharpshooter fires a rifle while standing with the butt of the gun against his shoulder. If the forward momentum of a bullet is the same as the backward momentum of the gun, why isn't it as dangerous to be hit by the gun as by the bullet?

Answer   It is the product mv which is the same for both the bullet and the gun. The bullet has a large velocity and a small mass, while the gun has a small velocity and a large mass. Furthermore, the bullet carries much more kinetic energy than the gun.

13. Does the center of mass of a rocket in free space accelerate? Explain. Can the speed of a rocket exceed the exhaust speed of the fuel? Explain.

Answer   The center of mass of a rocket, plus its exhaust does not accelerate: momentum must be conserved. However, if you consider the "rocket" to be the mechanical system plus the unexpended fuel, then it becomes obvious that the center of mass of that "rocket" does accelerate.

Ultimately, a rocket is no more complicated than two rubber balls that are pressed together, and released. One springs in one direction with one velocity; the other springs in the other direction with a velocity equal to

$$v_2 = v_1(m_1 / m_2)$$

In a similar manner, the mass of the rocket times its speed must equal the mass of all parts of the exhaust, times their speed. After a sufficient quantity of fuel has been exhausted the ratio m_1 / m_2 will be greater than 1; then the speed of the rocket **can** exceed the speed of the exhaust.

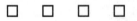

SOLUTIONS TO SELECTED END-OF-CHAPTER PROBLEMS

7. An estimated force-time curve for a baseball struck by a bat is shown in Figure P8.7. From this curve, determine (a) the impulse delivered to the ball, (b) the average force exerted on the ball, and (c) the peak force exerted on the ball.

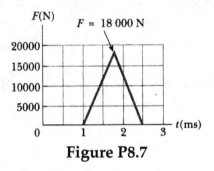

Figure P8.7

Solution The impulse delivered to the ball is equal to the area under the F-t graph. Thus,

(a) $I = \left(\dfrac{0 + 18000 \text{ N}}{2} \right)(2.5 \times 10^{-3} \text{ s} - 1.0 \times 10^{-3} \text{ s}) = 13.5 \text{ N} \cdot \text{s}$ ◊

(b) $\overline{F} = \dfrac{\int F \, dt}{\Delta t} = \dfrac{13.5 \text{ N} \cdot \text{s}}{(2.5 - 1.0)10^{-3} \text{ s}} = 9.00 \text{ kN}$ ◊

(c) From the graph, $F_{max} = 18000 \text{ N}$ ◊

---

9. A 3.00-kg steel ball strikes a wall with a speed of 10.0 m/s at an angle of 60.0° with the surface. It bounces off with the same speed and angle (Fig. P8.9). If the ball is in contact with the wall for 0.200 s, what is the average force exerted on the ball by the wall?

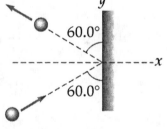

Figure P8.9

Solution We use the momentum version of the nonisolated system model, with the ball as our system:

$\Delta \mathbf{p} = \mathbf{F} \Delta t$:

$\Delta p_y = m\left(v_{fy} - v_{iy}\right) = m(v \cos 60.0°) - mv \cos 60.0° = 0$

$\Delta p_x = m\left(v_{fx} - v_{ix}\right) = m(-v \sin 60.0° - v \sin 60.0°) = -2mv \sin 60.0°$

$\Delta p_x = -2(3.00 \text{ kg})(10.0 \text{ m / s})(0.866) = -52.0 \text{ kg} \cdot \text{m / s}$

$\overline{\mathbf{F}} = \dfrac{\Delta \mathbf{p}}{\Delta t} = \dfrac{\Delta p_x \mathbf{i}}{\Delta t} = \dfrac{-52.0 \mathbf{i} \text{ kg} \cdot \text{m/s}}{0.200 \text{ s}} = -260 \mathbf{i} \text{ N}$ ◊

---

15. A 45.0-kg girl is standing on a plank that has a mass of 150 kg. The plank, originally at rest, is free to slide on a frozen lake, which is a flat, frictionless supporting surface. The girl begins to walk along the plank at a constant speed of 1.50 m/s relative to the plank. (a) What is her speed relative to the ice surface? (b) What is the speed of the plank relative to the ice surface?

Solution $\mathbf{v}_g$ = velocity of the girl relative to the ice

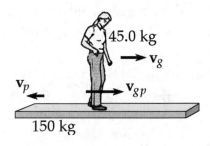
45.0 kg

$\mathbf{v}_{gp}$ = velocity of the girl relative to the plank

$\mathbf{v}_p$ = velocity of the plank relative to the ice

150 kg

The girl and the plank exert forces on each other, but the ice isolates them from outside horizontal forces. Therefore, the net momentum is zero for the combined girl plus plank system.

$$0 = m_g \mathbf{v}_g + m_p \mathbf{v}_p$$

Further, the relation among relative speeds can be written $\mathbf{v}_g = \mathbf{v}_{gp} + \mathbf{v}_p$:

$$\mathbf{v}_g = 1.50\mathbf{i} \ \text{m/s} + \mathbf{v}_p$$

We substitute: $0 = (45.0 \ \text{kg})(1.50\mathbf{i} \ \text{m/s} + \mathbf{v}_p) + (150 \ \text{kg})\mathbf{v}_p$

$$(195 \ \text{kg})\mathbf{v}_p = -(-45.0 \ \text{kg})(1.50\mathbf{i} \ \text{m/s})$$

(b) $\mathbf{v}_p = -0.346\mathbf{i} \ \text{m/s}$ ◊

(a) $\mathbf{v}_g = 1.50\mathbf{i} - 0.346\mathbf{i} \ \text{m/s} = 1.15\mathbf{i} \ \text{m/s}$ ◊

---

19. A neutron in a reactor makes an elastic head-on collision with the nucleus of a carbon atom initially at rest. (a) What fraction of the neutron's kinetic energy is transferred to the carbon nucleus? (b) If the initial kinetic energy of the neutron is 1.60×10^{-13} J, find its final kinetic energy and the kinetic energy of the carbon nucleus after the collision. (The mass of the carbon nucleus is about 12.0 times the mass of the neutron.)

Solution

(a) This is a perfectly elastic head-on collision, so we use Eq. 8.19: $v_{1i} - v_{2i} = -\left(v_{1f} - v_{2f}\right)$

Let object 1 be the neutron, and object 2 be the carbon nucleus, with $m_2 = 12m_1$.

Since $v_{2i} = 0$, $v_{2f} = v_{1i} + v_{1f}$

Now, by conservation of momentum, $m_1 v_{1i} + m_2 v_{2i} = m_1 v_{1f} + m_2 v_{2f}$

or $m_1 v_{1i} = m_1 v_{1f} + 12 m_1 v_{2f}$

Substituting our velocity equation, $v_{1i} = v_{1f} + 12(v_{1i} + v_{1f})$

We solve $-11 v_{1i} = 13\, v_{1f}$: $v_{1f} = -\left(\dfrac{11}{13}\right) v_{1i}$

and $v_{2f} = v_{1i} - \left(\dfrac{11}{13}\right) v_{1i} = \left(\dfrac{2}{13}\right) v_{1i}$

The neutron's original kinetic energy is $\dfrac{1}{2} m_1 v_{1i}^2$

The carbon's final kinetic energy is $\dfrac{1}{2} m_2 v_{2f}^2 = \dfrac{1}{2}(12 m_1)\left(\dfrac{2}{13}\right)^2 v_{1i}^2 = \left(\dfrac{48}{169}\right)\left(\dfrac{1}{2}\right) m_1 v_{1i}^2$

So, $\left(\dfrac{48}{169}\right) = 0.284$ or   28.4%   of the total energy is transferred.   ◊

(b) For the carbon nucleus, $K_{2f} = (0.284)(1.60 \times 10^{-13}\ \text{J}) = 45.4\,\text{fJ}$ ◊

The collision is perfectly elastic, so the neutron retains the rest of the energy,

$K_{1f} = (1.60 - 0.454) \times 10^{-13}\ \text{J} = 1.15 \times 10^{-13}\ \text{J} = 115\ \text{fJ}$ ◊

21. A 12.0-g wad of sticky clay is hurled horizontally at a 100-g wooden block initially at rest on a horizontal surface. The clay sticks to the block. After impact, the block slides 7.50 m before coming to rest. If the coefficient of friction between the block and the surface is 0.650, what was the speed of the clay immediately before impact?

Solution   We use the momentum version of the isolated system model to analyze the collision, and the energy version of the isolated system model to analyze the subsequent sliding process. The collision, for which figures (1) and (2) are before and after pictures, is **totally inelastic**, and momentum is conserved for the system of clay and block:

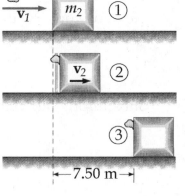

$$m_1 v_1 = (m_1 + m_2)v_2$$

In the sliding process occurring between figures (2) and (3), the original kinetic energy of the surface, block, and clay is equal to the increase in internal energy of the system due to friction:

$$\tfrac{1}{2}(m_1 + m_2)v_2{}^2 = f_f L$$

$$\tfrac{1}{2}(m_1 + m_2)v_2{}^2 = \mu(m_1 + m_2)gL$$

Solving for v_2:
$$v_2 = \sqrt{2\mu Lg} = \sqrt{2(0.650)(7.50 \text{ m})(9.80 \text{ m/s}^2)} = 9.77 \text{ m/s}$$

From the momentum conservation equation,

$$v_1 = \left(\frac{m_1 + m_2}{m_1}\right)v_2 = \left(\frac{112 \text{ g}}{12.0 \text{ g}}\right)9.77 \text{ m/s} = 91.2 \text{ m/s} \qquad \Diamond$$

---

27. A billiard ball moving at 5.00 m/s strikes a stationary ball of the same mass. After the collision, the first ball moves at 4.33 m/s, at an angle of 30.0° with respect to the original line of motion. Assuming an elastic collision (and ignoring friction and rotational motion), find the struck ball's velocity after the collision.

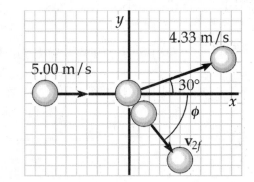

Solution

Call each mass m, and call $\mathbf{v}_{2f}$ the velocity of the second ball after the collision, as in the figure. Then apply conservation of momentum to the two-ball system.

In the x direction, $\qquad m(5.00 \text{ m/s}) = m(4.33 \text{ m/s}) \cos 30° + mv_{2fx} \qquad v_{2fx} = 1.25 \text{ m/s}$

In the y direction, $\qquad 0 = m(4.33 \text{ m/s}) \sin 30° + mv_{2fy} \qquad v_{2fy} = -2.17 \text{ m/s}$

$$\mathbf{v}_{2f} = 1.25\mathbf{i} - 2.17\mathbf{j} \quad (\text{or } 2.50 \text{ m/s at } -60.0°) \qquad \lozenge$$

We did not have to use the fact that the collision is elastic.

---

29. A 3.00-kg object with an initial velocity of $5.00\mathbf{i}$ m/s collides with and sticks to a 2.00-kg object with an initial velocity of $-3.00\mathbf{j}$ m/s. Find the final velocity of the composite mass.

Solution Momentum of the two-object system is conserved, with both objects having the same final velocity:

$$m_1\mathbf{v}_{1i} + m_2\mathbf{v}_{2i} = m_1\mathbf{v}_{1f} + m_2\mathbf{v}_{2f}$$

$$(3.00 \text{ kg})(5.00\mathbf{i} \text{ m/s}) + (2.00 \text{ kg})(-3.00\mathbf{j} \text{ m/s}) = (3.00 \text{ kg} + 2.00 \text{ kg})\mathbf{v}_f$$

$$\mathbf{v}_f = \frac{15.0\mathbf{i} - 6.00\mathbf{j}}{5.00} \text{ m/s} = (3.00\mathbf{i} - 1.20\mathbf{j}) \text{ m/s} \qquad \lozenge$$

Related Calculation: Compute the kinetic energy of the system both before and after the collision; show that kinetic energy is not conserved.

$$K_{1i} + K_{2i} = \frac{1}{2}(3.00 \text{ kg})(5.00 \text{ m/s})^2 + \frac{1}{2}(2.00 \text{ kg})(3.00 \text{ m/s})^2$$

$$K_{1i} + K_{2i} = 46.5 \text{ J}$$

$$K_{1f} + K_{2f} = \frac{1}{2}(5.00 \text{ kg})\left[(3.00 \text{ m/s})^2 + (1.20 \text{ m/s})^2\right] = 26.1 \text{ J} \qquad \lozenge$$

---

31. An unstable atomic nucleus of mass 17.0×10^{-27} kg initially at rest disintegrates into three particles. One of the particles, of mass 5.00×10^{-27} kg, moves along the y axis with a velocity of 6.00×10^6 m / s. Another particle, of mass 8.40×10^{-27} kg, moves along the x axis with a speed of 4.00×10^6 m / s. Find (a) the velocity of the third particle and (b) the total kinetic energy increase in the process.

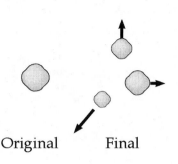

Original   Final

Solution

(a) With three particles, the total final momentum of the system is $m_1\mathbf{v}_{1f} + m_2\mathbf{v}_{2f} + m_3\mathbf{v}_{3f}$, and it must be zero to equal the original momentum.

The mass of the third particle is $m_3 = (17.0 - 5.00 - 8.40) \times 10^{-27}$ kg $= 3.60 \times 10^{-27}$ kg

The total momentum is zero: $m_1\mathbf{v}_{1f} + m_2\mathbf{v}_{2f} + m_3\mathbf{v}_{3f} = 0$

Solving for $\mathbf{v}_{3f}$, $\mathbf{v}_{3f} = -\dfrac{m_1\mathbf{v}_{1f} + m_2\mathbf{v}_{2f}}{m_3}$

$$\mathbf{v}_{3f} = -\frac{(3.00\mathbf{j} + 3.36\mathbf{i}) \times 10^{-20} \text{ kg} \cdot \text{m/s}}{3.60 \times 10^{-27} \text{ kg}} = (-9.33\mathbf{i} - 8.33\mathbf{j}) \text{ Mm/s} \qquad \lozenge$$

(b) The original kinetic energy of the system is zero.

The final kinetic energy is $K = K_{1f} + K_{2f} + K_{3f}$.

$$K_{1f} = \tfrac{1}{2}(5.00 \times 10^{-27} \text{ kg})(6.00 \times 10^6 \text{ m/s})^2 = 9.00 \times 10^{-14} \text{ J}$$

$$K_{2f} = \tfrac{1}{2}(8.40 \times 10^{-27} \text{ kg})(4.00 \times 10^6 \text{ m/s})^2 = 6.72 \times 10^{-14} \text{ J}$$

$$K_{3f} = \tfrac{1}{2}(3.60 \times 10^{-27} \text{ kg})\left((-9.33 \times 10^6 \text{ m/s})^2 + (-8.33 \times 10^6 \text{ m/s})^2\right) = 28.2 \times 10^{-14} \text{ J}$$

and $K = 9.00 \times 10^{-14} \text{ J} + 6.72 \times 10^{-14} \text{ J} + 28.2 \times 10^{-14} \text{ J} = 4.39 \times 10^{-13} \text{ J} = 439 \text{ fJ} \qquad \lozenge$

35. A uniform piece of sheet steel is shaped as in Figure P8.35. Compute the x and y coordinates of the center of mass of the piece.

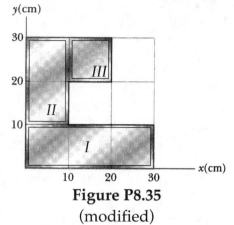

Figure P8.35
(modified)

Solution

Think of the sheet as composed of three sections, and consider the mass of each section to be at the geometric center of that section. Define the mass per unit area to be σ, and number the rectangles as shown. We can then calculate the mass and identify center of mass of each section.

$m_I = (30.0 \text{ cm})(10.0 \text{ cm})\sigma$ $\qquad$ $CM_I = (15.0 \text{ cm}, 5.0 \text{ cm})$

$m_{II} = (10.0 \text{ cm})(20.0 \text{ cm})\sigma$ $\qquad$ $CM_{II} = (5.0 \text{ cm}, 20.0 \text{ cm})$

$m_{III} = (10.0 \text{ cm})(10.0 \text{ cm})\sigma$ $\qquad$ $CM_{III} = (15.0 \text{ cm}, 25.0 \text{ cm})$

The overall CM is at a point defined by the vector equation $\mathbf{r}_{CM} \equiv \dfrac{\Sigma m_i \mathbf{r}_i}{\Sigma m_i}$

Substituting the appropriate values,

$$\mathbf{r}_{CM} = \frac{(300\sigma \text{ cm}^3)(15.0\mathbf{i} + 5.0\mathbf{j}) + (200\sigma \text{ cm}^3)(5.0\mathbf{i} + 20.0\mathbf{j}) + (100\sigma \text{ cm}^3)(15.0\mathbf{i} + 25.0\mathbf{j})}{(300 + 200 + 100)\sigma \text{ cm}^2}$$

$$\mathbf{r}_{CM} = \frac{(45.0\mathbf{i} + 15.0\mathbf{j} + 10.0\mathbf{i} + 40.0\mathbf{j} + 15.0\mathbf{i} + 25.0\mathbf{j})}{6.00} \text{ cm} = (11.7\mathbf{i} + 13.3\mathbf{j}) \text{ cm} \qquad \lozenge$$

If we chose any other division of the original shape, the answer would be the same.

---

37. A 2.00-kg particle has a velocity of $(2.00\mathbf{i} - 3.00\mathbf{j})$ m/s, and a 3.00-kg particle has a velocity of $(1.00\mathbf{i} + 6.00\mathbf{j})$ m/s. Find (a) the velocity of the center of mass and (b) the total momentum of the system of two particles.

Solution   Use $\mathbf{v}_{CM} = \dfrac{m_1\mathbf{v}_1 + m_2\mathbf{v}_2}{m_1 + m_2}$ and $\mathbf{p}_{CM} = (m_1 + m_2)\mathbf{v}_{CM}$:

(a) $\mathbf{v}_{CM} = \dfrac{(2.00\ \text{kg})[(2.00\mathbf{i} - 3.00\mathbf{j})\ \text{m / s}] + (3.00\ \text{kg})[(1.00\mathbf{i} + 6.00\mathbf{j})\ \text{m / s}]}{(2.00\ \text{kg} + 3.00\ \text{kg})}$

$\mathbf{v}_{CM} = (1.40\mathbf{i} + 2.40\mathbf{j})\ \text{m / s}$ ◊

(b) $\mathbf{p}_{CM} = (2.00\ \text{kg} + 3.00\ \text{kg})[(1.40\mathbf{i} + 2.40\mathbf{j})\ \text{m / s}] = (7.00\mathbf{i} + 12.0\mathbf{j})\ \text{kg} \cdot \text{m / s}$ ◊

Note that the center of mass is outside the boundary of the object, near the lower left-hand corner of the square it partially encloses.

---

39.   Romeo (77.0 kg) entertains Juliet (55.0 kg) by playing his guitar from the rear of their boat at rest in still water, 2.70 m away from Juliet in the front of the boat. After the serenade, Juliet carefully moves to the rear of the boat (away from shore) to plant a kiss on Romeo's cheek. How far does the 80.0-kg boat move toward the shore it is facing?

Solution   No outside forces act on the boat-plus-lovers system, so its momentum is conserved at zero and its center of mass stays fixed: $x_{CM, i} = x_{CM, f}$.

Define K to be the point where they kiss, and Δx_J and Δx_b as shown in the figure. Since Romeo moves with the boat (and thus $\Delta x_{Romeo} = \Delta x_b$), let m_b be the combined mass of Romeo and the boat.

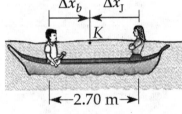

Then, $m_J \Delta x_J + m_b \Delta x_b = 0$

Choosing the x axis to point away from the shore, $(55.0\ \text{kg})\Delta x_J + (77.0\ \text{kg} + 80.0\ \text{kg})\Delta x_b = 0$

and $\Delta x_J = -2.85\ \Delta x_b$

As Juliet moves away from shore, the boat and Romeo glide toward the shore until the original 2.70 m gap between them is closed:

$$\Delta x_J - \Delta x_b = 2.70\ \text{m}$$

Substituting, we find $\Delta x_b = -0.700\ \text{m},\ \text{or}\ \ 0.700\ \text{m}\ \ \text{towards the shore}$ ◊

---

41. The first stage of a Saturn V space vehicle consumes fuel and oxidizer at the rate of 1.50×10^4 kg / s, with an exhaust speed of 2.60×10^3 m / s. (a) Calculate the thrust produced by these engines. (b) Find the acceleration of the vehicle just as it leaves the launch pad if its initial mass is 3.00×10^6 kg. [**Hint:** You must include the force of gravity to solve part (b).]

Solution

(a) The impulse due to the thrust, F, is equal to the change in momentum of the fuel as the fuel is exhausted from the rocket.

$$F = \frac{dp}{dt} = \frac{d}{dt}(mv_e)$$

Since the exhaust velocity v_e is constant as measured in the reference frame of the rocket,

$$F = v_e\left(\frac{dm}{dt}\right)$$

$\dfrac{dm}{dt}$ is the fuel consumption rate, so

$$F = \left(2.60 \times 10^3 \text{ m/s}\right)\left(1.50 \times 10^4 \text{ kg/s}\right)$$

$$F = 39.0 \text{ MN} \qquad \lozenge$$

(b) To find the acceleration, we use the particle under a net force model. Applying $\Sigma F = ma$,

$$\left(3.90 \times 10^7 \text{ N}\right) - \left(3.00 \times 10^6 \text{ kg}\right)\left(9.80 \text{ m / s}^2\right) = \left(3.00 \times 10^6 \text{ kg}\right)a$$

$$a = \frac{\left(3.90 \times 10^7 \text{ N}\right) - \left(29.4 \times 10^6 \text{ N}\right)}{3.00 \times 10^6 \text{ kg}} = 3.20 \text{ m/s}^2 \text{ up} \qquad \lozenge$$

Chapter 9

Relativity

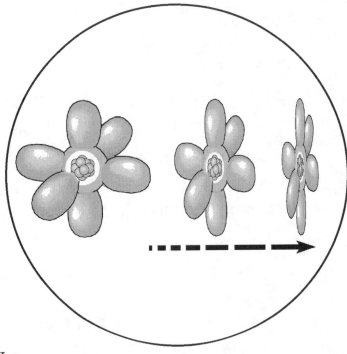

INTRODUCTION

Most of our everyday experiences and observations have to do with objects that move at speeds much less than that of light. Newtonian mechanics and early ideas on space and time were formulated to describe the motion of such objects. This formalism is very successful in describing a wide range of phenomena that occur at low speeds. It fails, however, when applied to particles whose speeds approach that of light. For example, it is possible to accelerate an electron to a speed of $0.99c$ (where c is the speed of light) by using a potential difference of several million volts. According to Newtonian mechanics, if the potential difference is increased by a factor of 4, the electron speed should jump to $1.98c$. However, experiments show that the speed of the electron always remains less than the speed of light, regardless of the size of the accelerating voltage. Thus, we can see that the validity of Newtonian mechanics is limited to low-velocity interactions.

In 1905, at the age of only 26, Einstein published his special theory of relativity. With this theory, experimental observations can be correctly predicted over the range of speeds from $v = 0$ to speeds approaching the speed of light. Newtonian mechanics, which was accepted for over 200 years, is in fact a special case of Einstein's theory. This chapter gives an introduction to the special theory of relativity, with emphasis on some of its consequences.

NOTES FROM SELECTED CHAPTER SECTIONS

9.1 The Principle of Newtonian Relativity

According to the principle of Newtonian relativity, the laws of mechanics are the same in all inertial frames of reference. Inertial frames of reference are those coordinate systems which are **at rest with respect to one another** or which move at constant velocity with respect to one another.

9.2 The Michelson-Morley Experiment

This experiment was designed to detect the velocity of the Earth with respect to the hypothetical **luminiferous ether**. The outcome of the experiment was **negative**, contradicting the ether hypothesis.

9.3 Einstein's Principle of Relativity

Einstein's special theory of relativity is based on two postulates:

- The laws of physics are the same in all **inertial** reference systems.
- The speed of light in vacuum is always measured to be 3.00×10^8 m/s, and the value is **independent** of the motion of the observer or of the motion of the source of light.

The second postulate is consistent with the negative results of the Michelson-Morley experiment which failed to detect the presence of an ether and suggested that the speed of light is the same in all inertial frames.

9.4 Consequences of Special Relativity

Two events that are simultaneous in one reference frame are, in general, not simultaneous in another frame which is moving with respect to the first.

The **proper time interval** is always the time interval measured by a clock at rest in the frame of reference of the measurement. According to an observer, a clock moving with respect to the observer runs slower than an identical clock at rest with respect to the observer. This effect is known as **time dilation.**

The **proper length** of an object is defined as the length of the object measured in the **reference frame in which the object is at rest.** The length of an object measured in a reference frame in which the object is moving is always less than the proper length. This effect is known as **length contraction.** The contraction occurs only **along the direction of motion.**

9.6 Relativistic Momentum and the Relativistic Form of Newton's Laws

To account for **relativistic effects**, it is necessary to modify the definition of momentum to satisfy the following conditions:

- The relativistic momentum must be conserved in all collisions.
- The relativistic momentum must approach the classical value, *mv*, as the quantity *v/c* approaches zero.

EQUATIONS AND CONCEPTS

In order to explain the motion of particles moving at speeds approaching the speed of light, one must use the **Lorentz coordinate transformation equations.** These equations represent the transformation between any two inertial frames in **relative** motion with velocity **v** in the *x* direction.

$$t' = \gamma\left(t - \frac{v}{c^2}x\right)$$

$$x' = \gamma(x - vt)$$

where

$$\gamma \equiv \frac{1}{\sqrt{1 - \frac{v^2}{c^2}}}$$

(9.8)

$$y' = y$$

$$z' = z$$

The **Lorentz velocity transformation equations** relate the observed velocity **u'** in the moving (S') frame to the measured velocity **u** in the S frame, and the relative velocity **v** (in the *x* direction) of S' with respect to S.

$$u'_x = \frac{u_x - v}{1 - \frac{u_x v}{c^2}}$$

(9.11)

$$u'_y = \frac{u_y}{\gamma\left(1 - \frac{u_x v}{c^2}\right)}$$

(9.12)

$$u'_z = \frac{u_z}{\gamma\left(1 - \frac{u_x v}{c^2}\right)}$$

Consider a single clock in the S' frame in which an observer in this frame measures the time interval of an event Δt_p (say the time it takes a light pulse to travel from a source to a mirror and back to the source). The time interval Δt for this event as measured by an observer in S is greater than Δt_p, since the observer in S must make clock readings at two different positions. That is, **a moving clock appears to run slower than an identical clock at rest with respect to the observer.** This is known as time dilation.

$$\Delta t = \frac{\Delta t_p}{\sqrt{1 - \dfrac{v^2}{c^2}}} = \gamma \Delta t_p \tag{9.6}$$

If an object moves along the x axis with a speed v, and has a length L_p along the direction of motion as measured in the moving frame, the length L as measured by a stationary observer is shorter than L_p by the factor $1/\gamma$. This is called **length contraction,** and the length L_p measured in the reference frame in which the object is at rest is called the **proper length.**

$$L = L_p \left(1 - \frac{v^2}{c^2}\right)^{1/2} \tag{9.7}$$

The fact that observers in the S and S' frames do not reach the same conclusions regarding such measurements can be understood by recognizing that simultaneity is not an absolute concept. That is, **Two events that are simultaneous in one reference frame are not simultaneous in another reference frame that is in motion relative to the first.**

The relativistic equation for the **momentum** of a particle of mass m moving with a speed u satisfies the conditions that (1) momentum is conserved in all collisions and (2) p approaches the classical expression as u approaches 0. That is, as $u \Rightarrow 0, p \Rightarrow mu$.

$$\mathbf{p} \equiv \frac{m\mathbf{u}}{\sqrt{1 - \frac{u^2}{c^2}}}$$

(9.14)

In the relativistic equation for the **kinetic energy** of a particle of mass m moving with a speed u, the term mc^2 is called the **rest energy.**

$$K = \frac{mc^2}{\sqrt{1 - \frac{u^2}{c^2}}} - mc^2 = (\gamma - 1)mc^2$$

(9.18)

The **total energy** E of a particle is the sum of the kinetic energy and the rest energy.

$$E = \gamma mc^2 = K + mc^2 = K + E_R$$

(9.20)

The expression for the total energy can be written when γ is replaced with its equivalent. This expression shows that **mass and energy are equivalent concepts**; that is, mass is a form of energy.

$$E = \frac{mc^2}{\sqrt{1 - \frac{u^2}{c^2}}}$$

(9.21)

When the momentum or energy of a particle is known (rather than the speed), it is useful to have an expression relating the total energy and relativistic momentum.

$$E^2 = p^2c^2 + (mc^2)^2$$

(9.22)

There is an exact expression relating energy and momentum for particles which have zero mass (e.g. photons). These zero-mass particles always travel at the speed of light.

$$E = pc \qquad (9.23)$$

A convenient energy unit to use to express the energies of electrons and other subatomic particles is the electron volt (eV).

$$1 \text{ eV} = 1.60 \times 10^{-19} \text{ J}$$

REVIEW CHECKLIST

▷ State Einstein's two postulates of the special theory of relativity.

▷ Understand the Michelson-Morley experiment, its objectives, results, and the significance of its outcome.

▷ Understand the idea of simultaneity, and the fact that simultaneity is not an absolute concept. That is, two events which are simultaneous in one reference frame are not simultaneous when viewed from a second frame moving with respect to the first.

▷ Make calculations using the equations for time dilation, relativistic momentum, and length contraction.

▷ State the correct relativistic expressions for the momentum, kinetic energy, and total energy of a particle. Make calculations using these equations.

ANSWERS TO SELECTED CONCEPTUAL QUESTIONS

6. Explain why it is necessary, when defining length, to specify that the positions of the ends of a rod are to be measured simultaneously.

Answer Suppose a railroad train is moving past you. One way to measure its length is this: You mark the tracks at the front of the moving engine at 9:00:00 AM, while your assistant marks the tracks at the back of the caboose at the same time. Then you find the distance between the marks on the tracks with a tape measure. You and your assistant must make the marks simultaneously (in your reference frame), for otherwise the motion of the train would make its length different from the distance between marks.

8. Give a physical argument that shows that it is impossible to accelerate an object of mass m to the speed of light, even with a continuous force acting on it.

Answer As an object approaches the speed of light, its energy approaches infinity. Hence, it would take an infinite amount of work to accelerate the object to the speed of light under the action of a continuous force or it would take an infinitely large force.

9. List some ways our everyday lives would change if the speed of light were only 50 m/s.

Answer For a wonderful fictional exploration of this question, get a "Mr. Tompkins" book by George Gamow. All of the relativity effects would be obvious in our lives. Time dilation and length contraction would both occur. Driving home in a hurry, you would push on the gas pedal not to increase your speed very much, but to make the blocks shorter. Big Doppler shifts in wave frequencies would make red lights look green as you approached, and make car horns and radios useless. High-speed transportation would be both very expensive, requiring huge fuel purchases, as well as dangerous, since a speeding car could knock down a building. When you got home, hungry for lunch, you would find that you had missed dinner; there would be a five-day delay in transit when you watched the Olympics in Australia on live TV. Finally, we would not be able to see the Milky Way, since the fireball of the Big Bang would surround us at the distance of Rigel, or Deneb.

SOLUTIONS TO SELECTED END-OF-CHAPTER PROBLEMS

3. In a laboratory frame of reference, an observer notes that Newton's second law is valid. Show that it is also valid for an observer moving at a constant speed, small compared to the speed of light, relative to the laboratory frame.

Solution The first observer watches some object accelerate along the x direction under applied forces. Call the instantaneous velocity of the object u_x. The second observer moves along the x axis at constant speed v relative to the first, and measures the object to have velocity

$$u'_x = u_x - v$$

The acceleration is $\qquad \dfrac{du'_x}{dt} = \dfrac{du_x}{dt} - 0$

This is the same as that measured by the first observer. In this nonrelativistic case, they measure also the same mass and forces; so the second observer also confirms that $\Sigma F = ma$. ◊

---

9. An atomic clock moves at 1000 km/h for 1.00 h as measured by an identical clock fixed to the Earth. How many nanoseconds slow will the moving clock be at the end of the one-hour interval?

Solution This problem is slightly more difficult than most, for the simple reason that your calculator probably cannot hold enough decimal places to yield an accurate answer. However, we can bypass the difficulty by noting the approximation:

$$\sqrt{1 - \frac{v^2}{c^2}} \cong 1 - \frac{v^2}{2c^2}$$

Squaring both sides will show that when v/c is small, these two terms are equivalent.

Solving for v/c, $\qquad \dfrac{v}{c} = \left(\dfrac{1000 \times 10^3 \text{ m / h}}{3.00 \times 10^8 \text{ m / s}} \right)\left(\dfrac{1 \text{ h}}{3600 \text{ s}} \right) = 9.26 \times 10^{-7}$

From Equation 9.6,
$$\Delta t = \gamma \Delta t_p = \frac{\Delta t_p}{\sqrt{1 - v^2/c^2}}$$

Rearranging, our approximation yields
$$\Delta t_p = \left(\sqrt{1 - \frac{v^2}{c^2}}\right)\Delta t \cong \left(1 - \frac{v^2}{2c^2}\right)\Delta t$$

and
$$\Delta t - \Delta t_p = \frac{v^2}{2c^2}\Delta t$$

Substituting,
$$\Delta t - \Delta t_p = \frac{\left(9.26 \times 10^{-7}\right)^2}{2}(3600\ \text{s})$$

Thus the time lag of the moving clock is $\Delta t - \Delta t_p = 1.54 \times 10^{-9}$ s $= 1.54$ ns   ◊

---

11. A spaceship with a proper length of 300 m takes 0.750 μs seconds to pass an Earth observer. Determine its speed as measured by the Earth observer.

Solution

We should first determine if the spaceship is traveling at a relativistic speed: classically, $v = (300\text{m})/(0.750\ \mu s) = 4.00 \times 10^8$ m/s, which is faster than the speed of light (impossible)! Quite clearly, the relativistic correction must be used to find the correct speed of the spaceship, which we can guess will be close to the speed of light.

We can use the contracted length equation to find the speed of the spaceship in terms of the proper length and the time. The time of 0.750 μs is the **proper time** measured by the Earth observer, because it is the time interval between two events that she sees as happening at the same point in space. The two events are the passage of the front end of the spaceship over her stopwatch, and the passage of the back end of the ship.

$L = L_p / \gamma$, with $L = v\Delta t$: $v\Delta t = L_p\left(1 - v^2/c^2\right)^{1/2}$

Squaring both sides, $\qquad v^2 \Delta t^2 = L_p^{\,2}\left(1 - v^2 / c^2\right)$

$$v^2 c^2 = L_p^{\,2} c^2 / \Delta t^2 - v^2 L_p^{\,2} / \Delta t^2$$

Solving for the speed, $\qquad v = \dfrac{c L_p / \Delta t}{\sqrt{c^2 + L_p^{\,2} / \Delta t^2}}$

So $\qquad v = \dfrac{\left(3.00 \times 10^8\right)(300 \text{ m}) \big/ \left(0.750 \times 10^{-6} \text{ s}\right)}{\sqrt{\left(3.00 \times 10^8\right)^2 + (300 \text{ m})^2 \big/ \left(0.750 \times 10^{-6} \text{ s}\right)^2}} = 2.40 \times 10^8 \text{ m / s } \Diamond$

The spaceship is traveling at $0.8c$. We can also verify that the general equation for the speed reduces to the classical relation $v = L_p / \Delta t$ when the time is relatively large.

---

19.   Two jets of material from the center of a radio galaxy are ejected in opposite directions. Both jets move at $0.750c$ relative to the galaxy. Determine the speed of one jet relative to the other.

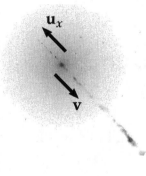

Solution

Take the galaxy as the unmoving frame. Arbitrarily define the jet moving upwards to be the object, and the jet moving downwards to be the "moving" frame.

$$u_x = 0.750 \; c \qquad\qquad\qquad v = -0.750 \; c$$

Thus $\qquad u_x' = \dfrac{u_x - v}{1 - u_x v/c^2} = \dfrac{0.750c - (-0.750c)}{1 - (0.750c)(-0.750c)/c^2} = \dfrac{1.50c}{1 + 0.750^2} = 0.960c \qquad\qquad \Diamond$

---

27. An unstable particle at rest breaks into two fragments of **unequal** mass. The mass of one fragment is 2.50×10^{-28} kg, and that of the other is 1.67×10^{-27} kg. If the lighter fragment has a speed of $0.893c$ after the breakup, what is the speed of the heavier fragment?

Solution   We use the momentum version of the isolated system model.

Refer to the heavier particle with the subscript '2', and to the lighter particle with the subscript '1.'

Relativistic momentum of the system of two particles must be conserved. For the total momentum to be zero after the fission (breakup), as it was before, $p_2 + p_1 = 0$:

$$\gamma_2 m_2 v_2 + \gamma_1 m_1 v_1 = 0$$

$$\gamma_2 m_2 v_2 + \left(\frac{2.50 \times 10^{-28} \text{ kg}}{\sqrt{1 - 0.893^2}} \right)(0.893c) = 0$$

Rearranging,

$$\left(\frac{1.67 \times 10^{-27} \text{ kg}}{\sqrt{1 - v_2^2/c^2}} \right)\frac{v_2}{c} = -4.96 \times 10^{-28} \text{ kg}$$

Squaring both sides,

$$\left(2.79 \times 10^{-54} \text{ kg}^2 \right)\left(\frac{v_2}{c} \right)^2 = \left(2.46 \times 10^{-55} \text{ kg}^2 \right)\left(1 - \frac{v_2^2}{c^2} \right)$$

and

$$v_2 = -0.285c$$

The negative sign only means that the two particles must move in **opposite** directions. The speed, then is

$$\left| v_2 \right| = 0.285c \qquad \diamondsuit$$

31. A proton moves at 0.950c. Calculate its (a) rest energy, (b) total energy, and (c) kinetic energy.

Solution At $v = 0.950c$, $\gamma = 3.20$

(a) $E_R = mc^2 = (1.67 \times 10^{-27} \text{ kg})(2.998 \times 10^8 \text{ m/s})^2 = 1.50 \times 10^{-10} \text{ J} = 938 \text{ MeV}$ ◊

(b) $E = \gamma mc^2 = \gamma E_R = (3.20)(938 \text{ MeV}) = 3.00 \text{ GeV}$ ◊

(c) $K = E - E_R = 3.00 \text{ GeV} - 938 \text{ MeV} = 2.07 \text{ GeV}$ ◊

---

35. Show that the energy-momentum relationship $E^2 = p^2c^2 + (mc^2)^2$ follows from the expressions $E = \gamma mc^2$ and $p = \gamma mu$.

Solution
$$E = \gamma mc^2 \qquad p = \gamma mu$$

Squaring both equations,
$$E^2 = \left(\gamma mc^2\right)^2 \quad p^2 = \left(\gamma mu\right)^2$$

Multiplying the second equation by c^2, and subtracting it from the first,
$$E^2 - p^2c^2 = (\gamma mc^2)^2 - (\gamma mu)^2 c^2$$

$$E^2 - p^2c^2 = \gamma^2\left(\left(mc^2\right)\left(mc^2\right) - \left(mc^2\right)\left(mu^2\right)\right)$$

Extracting the $\left(mc^2\right)$ terms,
$$E^2 - p^2c^2 = \gamma^2\left(mc^2\right)^2\left(1 - \frac{u^2}{c^2}\right)$$

and
$$E^2 - p^2c^2 = \left(1 - \frac{u^2}{c^2}\right)^{-1}\left(mc^2\right)^2\left(1 - \frac{u^2}{c^2}\right)$$

Thus,
$$E^2 - p^2c^2 = \left(mc^2\right)^2$$ ◊

---

37. A pion at rest ($m_\pi = 270m_e$) decays to a muon ($m_\mu = 207m_e$) and an antineutrino ($m_{\bar{\nu}} \cong 0$). The reaction is written $\pi^- \to \mu^- + \bar{\nu}$. Find the kinetic energy of the muon and the energy of the antineutrino in electron volts. (**Hint:** Relativistic momentum is conserved.)

Solution We use, together, both the energy version and the momentum version of the isolated system model.

By conservation of energy,
$$m_\pi c^2 = \gamma m_\mu c^2 + |p_\nu|c$$

By conservation of momentum,
$$p_\nu = -p_\mu = -\gamma m_\mu v$$

Substituting the second equation into the first,
$$m_\pi c^2 = \gamma m_\mu c^2 + \gamma m_\mu vc$$

Simplified, this equation then reads
$$m_\pi = m_\mu \left(\gamma + \gamma v / c \right)$$

Substituting for the masses,
$$270 m_e = (207 m_e)\left(\gamma + \frac{\gamma v}{c} \right)$$

where
$$m_e c^2 = 0.511 \text{ MeV}.$$

Numerically,
$$\frac{270 m_e}{207 m_e} = 1.30 = \frac{1 + v/c}{\sqrt{1 - (v/c)^2}}$$

Solving for v/c with the quadratic formula,
$$\frac{v}{c} = \frac{270^2 - 207^2}{270^2 + 207^2} = 0.260$$

Therefore,
$$\gamma = \frac{1}{\sqrt{1 - v^2 / c^2}} = 1.0355$$

and
$$K_\mu = (0.0355)(207 \times 0.511 \text{ MeV}) = 3.76 \text{ MeV} \qquad \lozenge$$

$$K_{\bar{\nu}} = (270 \times 0.511 \text{ MeV}) - (207 \times 0.511 \text{ MeV} + 3.76 \text{ MeV}) = 28.4 \text{ MeV} \qquad \lozenge$$

39. The power output of the Sun is 3.77×10^{26} W. How much mass is converted to energy in the Sun each second?

Solution From $E_R = mc^2$, we have

$$m = \frac{E_R}{c^2} = \frac{3.77 \times 10^{26} \text{ J}}{(3.00 \times 10^8 \text{ m / s})^2} = 4.19 \times 10^9 \text{ kg}$$ ◊

45. The cosmic rays of highest energy are protons that have kinetic energy on the order of 10^{13} MeV. (a) How long would it take a proton of this energy to travel across the Milky Way galaxy, having a diameter $\sim 10^5$ light-years, as measured in the proton's frame? (b) From the point of view of the proton, how many kilometers across is the galaxy?

Solution

We first find the proton's speed in the frame of reference where the galaxy is stationary:

For a proton, $mc^2 = (1.67 \times 10^{-27} \text{ kg})(3.00 \times 10^8 \text{ m / s})^2 \left(\frac{1 \text{ eV}}{1.60 \times 10^{-19} \text{ kg} \cdot \text{m}^2 / \text{s}^2} \right) = 938 \text{ MeV}$

Now, $K = (\gamma - 1)(938 \text{ MeV})$

$10^{13} \text{ MeV} = (\gamma - 1)(938 \text{ MeV})$ and $\gamma = 1.07 \times 10^{10} = 1/\sqrt{1 - v^2 / c^2}$

$1 - v^2 / c^2 = 8.80 \times 10^{-21}$

$v = c\sqrt{1 - 8.80 \times 10^{-21}}$

This speed is nearly as large as the speed of light. In the galaxy frame the traversal time is

$\Delta t = x / v = 10^5$ light years$/ c = 10^5$ years

(a) This is dilated from the proper time measured in the proton's frame: $\Delta t = \gamma \Delta t'$

$\Delta t' = \Delta t / \gamma = 10^5 \text{ yr} / 1.07 \times 10^{10} = 9.38 \times 10^{-6}$ years = 296 s

$\Delta t' \sim$ A few hundred seconds ◊

(b) The proton sees the galaxy moving across the proton, at speed nearly equal to c, in 296 s:

$\Delta L' = (3.00 \times 10^8)(296 \text{ s}) = 8.88 \times 10^7 \text{ km} \sim 10^8 \text{ km}$ ◊

The observers on the proton and on the galaxy agree on the value v of their speed of relative motion and on the value of γ.

---

51. A supertrain (proper length 100 m) travels at a speed of $0.950c$ as it passes through a tunnel (proper length 50.0 m). As seen by a trackside observer, is the train ever completely within the tunnel? If so, with how much space to spare?

Solution

The observer sees the proper length of the tunnel, 50.0 m, but sees the train Lorentz-contracted to length

$L = L_p \sqrt{1 - v^2/c^2} = (100 \text{ m})\sqrt{1 - (0.950c)^2} = 31.2 \text{ m}$

This is shorter than the tunnel by 18.8 m, so it is completely within the tunnel. ◊

---

53. A particle with electric charge q moves along a straight line in a uniform electric field $\mathbf{E}$ with a speed of u. The electric force exerted on the particle is $q\mathbf{E}$. If the motion and the electric field are both in the x direction, (a) show that the acceleration of the charge q in the x direction is given by

$$a = \frac{du}{dt} = \frac{qE}{m}\left(1 - \frac{u^2}{c^2}\right)^{3/2}$$

(b) Discuss the significance of the dependence of the acceleration on the speed. (c) If the particle starts from rest at $x = 0$ at $t = 0$, how would you proceed to find the speed of the particle and its position at time t?

Solution   As you will learn later, the force on a charge in an electric field is given by $F = qE$. Further, at any speed, the **momentum** of the particle is given by

$$p = \gamma mu = \frac{mu}{\sqrt{1 - u^2/c^2}}$$

(a)   We use the momentum version of the nonisolated system model.

Since $F = qE = \dfrac{dp}{dt}$,

$$qE = \frac{d}{dt}\left(mu\left(1 - u^2/c^2\right)^{-1/2}\right)$$

$$qE = m\left(1 - \frac{u^2}{c^2}\right)^{-1/2}\frac{du}{dt} + \tfrac{1}{2}mu\left(1 - \frac{u^2}{c^2}\right)^{-3/2}\left(\frac{2u}{c^2}\right)\frac{du}{dt}$$

Simplifying, we find that $\dfrac{qE}{m} = \dfrac{du}{dt}\left(1 - \dfrac{u^2}{c^2}\right)^{-3/2}$ and $a = \dfrac{du}{dt} = \dfrac{qE}{m}\left(1 - \dfrac{u^2}{c^2}\right)^{3/2}$ ◊

(b)   As $u \to c$, we see that $a \to 0$. The particle thus never attains the speed of light.

(c)   Taking the acceleration equation, isolating the velocity terms and integrating,

$$\int_0^u \left(1 - \frac{u^2}{c^2}\right)^{-3/2} du = \int_0^t \frac{qE}{m}\,dt \qquad u = \frac{qEct}{\sqrt{m^2c^2 + q^2E^2t^2}} = \frac{dx}{dt}$$

$$x = \int_0^x dx = qEc\int_0^t \frac{t\,dt}{\sqrt{m^2c^2 + q^2E^2t^2}} = \frac{c}{qE}\left(\sqrt{m^2c^2 + q^2E^2t^2} - mc\right) \qquad ◊$$

Chapter 10

Rotational Motion

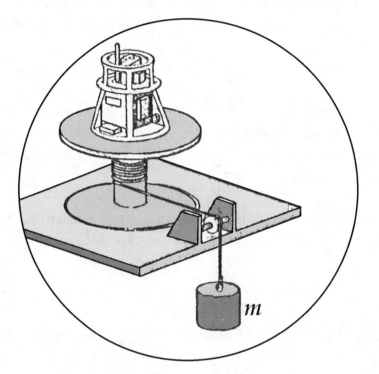

INTRODUCTION

When an extended object, such as a door on hinges, rotates about its axis, the motion cannot be analyzed by treating the object as a particle, since at any given time different parts of the object have different velocities and accelerations. For this reason, it is convenient to consider an extended object as a large number of particles, each with its own velocity and acceleration.

In dealing with the rotation of an object, analysis is greatly simplified by assuming the object to be rigid. A rigid object is defined as one that is nondeformable or, to say the same thing another way, one in which the distances between all pairs of particles remain constant. In this chapter, we treat the rotation of a rigid object about a fixed axis, commonly referred to as **pure rotational motion**.

Part of this chapter is concerned with the conditions under which a rigid object is in equilibrium. The term **equilibrium** implies either that the object is at rest or that its center of mass moves with constant velocity. We deal here with the former, which is referred to as an object in **static equilibrium**.

In Chapter 4 we stated that one necessary condition for equilibrium is that the net force on an object be zero. If the object is treated as a particle, this is the only condition that must be satisfied for equilibrium. That is, if the net force on the particle is zero, the particle remains at rest (if originally at rest) or moves with constant velocity (if originally in motion.)

The situation with real (extended) objects is more complex because objects cannot be treated as particles. In order for an extended object to be in static equilibrium, the net force on it must be zero **and** it must have no tendency to rotate. This second condition of equilibrium requires that **the net torque about any origin be zero**. In order to establish whether or not an object is in equilibrium, we must know its size and shape, the forces acting on different parts of it, and the points of application of the various forces.

NOTES FROM SELECTED CHAPTER SECTIONS

10.1 Angular Speed and Angular Acceleration

Pure rotational motion refers to the motion of a rigid body about a fixed axis.

In the case of **rotation about a fixed axis**, every particle on the rigid body has the same angular velocity and the same angular acceleration.

One **radian** (rad) is the angle subtended by an arc length equal to the radius of the arc.

The angular position (θ), angular velocity (ω), and angular acceleration (α) are analogous to linear position (x), linear velocity (v), and linear acceleration (a), respectively. The variables, θ, ω, and α, differ dimensionally from the variables x, v, and a, only by a length factor.

10.2 Rotational Kinematics: the Rigid Body Under Constant Angular Acceleration

The **kinematic expressions** for rotational motion under constant angular acceleration are of the **same form** as those for linear motion under constant linear acceleration with the substitutions $x \to \theta$, $v \to \omega$, and $a \to \alpha$.

10.3 Relationships Between Rotational and Translational Quantities

When a rigid body rotates about a fixed axis, every part of the body has the same angular velocity and the same angular acceleration. However, different parts of the body, in general, have different linear velocities and different linear accelerations.

10.4 Rotational Kinetic Energy

From the definition of moment of inertia, we see that it has dimensions of ML^2 (kg·m² in SI units). It plays the role of mass in **all** rotational equations. Although we shall commonly refer to the quantity $\frac{1}{2}I\omega^2$ as the rotational kinetic energy, it is not a new form of energy. It is ordinary kinetic energy. It is important to recognize the analogy between kinetic energy associated with linear motion, $\frac{1}{2}mv^2$, and rotational kinetic energy, $\frac{1}{2}I\omega^2$. The quantities I and ω in rotational motion are analogous to m and v in linear motion, respectively.

10.5 Torque and the Vector Product

Torque is a physical quantity which is the measure of the tendency of a force to cause rotation of a body about a specified axis. It is important to remember that torque must be defined with respect to a **specific axis** of rotation. Torque, which has the **SI units** of N·m, must not be confused with force.

10.6 The Rigid Body in Equilibrium

A **rigid body** is defined as one that does not deform under the application of external forces. There are two necessary conditions for **equilibrium of a rigid body**: (1) the resultant external force must be zero and (2) the resultant external torque must be zero about any axis. Two forces applied to a rigid body will result in equilibrium if and only if the forces have equal magnitudes and opposite directions and the torques resulting from the forces on the body are equal in magnitude and opposite in direction.

10.8 Angular Momentum

The **torque** acting on a particle is equal to the time rate of change of its angular momentum.

10.9 Conservation of Angular Momentum

The **total angular momentum** of a system is constant if the net external torque acting on the system is zero. The resultant torque acting on an object about the center of mass of the object equals the time rate of change of angular momentum of the object, regardless of the motion of the center of mass.

10.11 Rolling of Rigid Bodies

Although the points on a rigid body rotating about a fixed axis may not experience the same force, linear acceleration, or linear velocity, every point on the body has the same angular acceleration and angular velocity at any instant. Therefore, at any instant the rigid body as a whole is characterized by specific values for angular acceleration, net torque, and angular velocity.

The **work-kinetic energy theorem in rotational motion** states that the net work done by external forces in rotating a rigid body about a fixed axis equals the change in the body's rotational kinetic energy.

The **total kinetic energy** of a body undergoing rolling motion is the sum of the rotational kinetic energy about the center of mass and the translational kinetic energy of the center of mass.

EQUATIONS AND CONCEPTS

When a particle moves along a circular path of radius r, the distance traveled by the particle is called the arc length, s. The radial line from the center of the circle to the particle sweeps out an angle, θ.

$$\theta = \frac{s}{r} \tag{10.1b}$$

The angle θ is the ratio of two lengths (arc length to radius) and hence is a dimensionless quantity. However, it is common practice to refer to the angle as being in units of radians. In calculations, the relationship between radians and degrees is the following:

$$\theta^{rad} = \left(\frac{2\pi}{360°}\right)\theta^{deg} = \left(\frac{\pi}{180°}\right)\theta^{deg}$$

The **average angular speed** $\overline{\omega}$ of a particle or body rotating about a fixed axis equals the ratio of the angular displacement $\Delta\theta$ to the time interval Δt, where θ is measured in radians.

$$\overline{\omega} = \frac{\Delta\theta}{\Delta t} \qquad (10.2)$$

The **instantaneous angular speed** ω is defined as the limit of the average angular velocity as Δt approaches zero.

$$\omega = \frac{d\theta}{dt} \qquad (10.3)$$

The **average angular acceleration** $\overline{\alpha}$ of a rotating body is defined as the ratio of the change in angular velocity to the time interval Δt.

$$\overline{\alpha} = \frac{\Delta\omega}{\Delta t} \qquad (10.4)$$

The **instantaneous angular acceleration** equals the limit of the average angular acceleration as Δt approaches zero.

$$\alpha = \frac{d\omega}{dt} \qquad (10.5)$$

If a particle or body rotates about a fixed axis with **constant** angular acceleration, we can apply the **equations of rotational kinematics.**

See Table 10.1 of the text for a comparison of equations for rotational and translational motion.

$$\omega_f = \omega_i + \alpha t \qquad (10.6)$$

$$\theta_f = \theta_i + \omega_i t + \frac{1}{2}\alpha t^2 \qquad (10.7)$$

$$\omega_f^2 = \omega_i^2 + 2\alpha(\theta_f - \theta_i) \qquad (10.8)$$

$$\theta_f = \theta_i + \frac{1}{2}(\omega_i + \omega_f)t \qquad (10.9)$$

If a rigid body rotates about a fixed axis, the linear speed of any point on the body a distance r from the axis of rotation is related to the angular speed through the relation $v = r\omega$. Similarly, the tangential acceleration of any point on the body is related to the angular acceleration through the relation $a_t = r\alpha$. Note that **every point on the body has the same ω and α, but not every point has the same v and a_t.**

$$v = r\omega \qquad (10.10)$$

$$a_t = r\alpha \qquad (10.11)$$

Remember that a particle moving in a circular path also experiences a centripetal acceleration.

$$a_c = \frac{v^2}{r} = r\omega^2 \qquad (10.12)$$

The moment of inertia I is a measure of a system's resistance to change its angular speed.

$$I = \sum_i m_i r_i^2 \quad \text{(system of particles)} \quad (10.14)$$

$$I = \int r^2 dm \quad \text{(continuous object)} \quad (10.16)$$

The **torque** τ due to an applied force has a magnitude given by the product of the force and its moment arm d, where d equals the **perpendicular distance** from the rotation axis to the line of action of **F**. Torque is a measure of the ability of a force to rotate a body about a specified axis. Note that the torque depends on the axis of rotation, which must be specified when τ is evaluated.

$$\tau \equiv rF\sin\phi \qquad (10.18)$$

The **torque** acting on a particle whose vector position is **r** can be expressed as **r** × **F**, where **F** is the external force acting on the particle. Torque also depends on the choice of the origin and has the SI unit of N·m.

$$\boldsymbol{\tau} \equiv \mathbf{r} \times \mathbf{F} \qquad (10.19)$$

The **cross product** of any two vectors **A** and **B** is a vector **C** whose magnitude is given by $AB\sin\theta$ and whose direction is perpendicular to the plane formed by **A** and **B**. The sense of **C** can be determined from the right-hand rule. See Section 10.5 of the text for properties of the unit vectors.

$$\mathbf{C} = \mathbf{A} \times \mathbf{B} \qquad (10.20)$$

$$|\mathbf{C}| \equiv AB\sin\theta \qquad (10.21)$$

In general, **an object at rest or one moving with constant velocity will only do so if the resultant force on it is zero.** This is a statement of the **first condition of equilibrium**, and corresponds to the condition of translational equilibrium.

$$\Sigma\mathbf{F} = 0 \qquad (10.24)$$

Since the left hand side of Equation 10.24 is a **vector sum** of all **external forces** acting on the body, this necessarily implies that the sum of the x, y, and z components separately must be zero.

$$\Sigma F_x = 0$$

$$\Sigma F_y = 0$$

$$\Sigma F_z = 0$$

The **second condition of equilibrium** of a rigid body requires that the **vector sum of the torques relative to any origin must be zero.** This is the condition of rotational equilibrium.

$$\Sigma\tau = 0 \qquad (10.25)$$

If all the forces acting on a rigid body lie in a common plane, say the x-y plane, then there is no z component of force. In this case, we only have to deal with three equations--two of which correspond to the first condition of equilibrium, the third coming from the second condition. In this case, the torque vector lies along the z axis. All problems in this chapter fall into this category.

$$\Sigma F_x = 0 \qquad (10.26)$$

$$\Sigma F_y = 0$$

$$\Sigma\tau_z = 0$$

The **net torque** acting on a rigid body about some axis is equal to the product of the moment of inertia and angular acceleration, where I is the moment of inertia about the fixed axis of rotation.

$$\sum \tau = I\alpha \qquad (10.27)$$

The work-kinetic energy theorem for pure rotation is given by Equation 10.29.

$$W = \frac{1}{2}I\omega_f^2 - \frac{1}{2}I\omega_i^2 \qquad (10.29)$$

Given a chosen rotation axis and an origin on the rotation axis, the **angular momentum** of a particle with linear momentum $\mathbf{p}$ and vector position $\mathbf{r}$ (relative to the origin) is defined as $\mathbf{L} = \mathbf{r} \times \mathbf{p}$. The SI unit of angular momentum is kg·m/s². Note that both the magnitude and direction of $\mathbf{L}$ depend on the choice of origin.

$$\mathbf{L} \equiv \mathbf{r} \times \mathbf{p} \qquad (10.32)$$

The **torque** on a particle about a chosen rotation axis **equals the time rate of change of its angular momentum** about the same axis. This expression is the rotational analog of Newton's second law, $\mathbf{F} = d\mathbf{p}/dt$, and is the basic equation for treating a particle moving in a path around some fixed point.

$$\tau = \frac{d\mathbf{L}}{dt} \qquad (10.36)$$

The angular momentum of a system of particles is obtained by taking the vector sum of the individual angular momenta about some point in an inertial frame. The individual momenta may change with time, which can change the total angular momentum. However, the total angular momentum of the system will only change if a net **external** torque acts on the system. In fact, **the net torque acting on a system of particles equals the time rate of change of the total angular momentum.**

$$\Sigma \tau_{ext} = \frac{d\mathbf{L}_{tot}}{dt} \qquad (10.37)$$

The **magnitude of the angular momentum of a rigid body** rotating in the x-y plane about a **fixed axis** (the z axis) is given by the product $I\omega$, where I is the moment of inertia about the axis of rotation and ω is the angular speed.

$$L_z = I\omega$$

The **law of conservation of angular momentum** states that if the resultant external torque acting on a system is zero, the total angular momentum of the system is constant. This follows from Equation 10.31.

$$\Sigma \tau_{ext} = \frac{d\mathbf{L}}{dt} = 0 \qquad (10.38)$$

$$\mathbf{L}_{tot} = \text{constant} \qquad (10.39)$$

If a zero net torque acts on a body rotating about a fixed axis, and the moment of inertia changes from I_i to I_f, then conservation of angular momentum can be used to find the final angular speed in terms of the initial angular speed.

$$I_i \omega_i = I_f \omega_f = \text{constant}$$

If a net torque τ acts on a rigid body, the **power supplied to the body** at any instant is proportional to the angular speed.

$$\mathcal{P} = \tau\omega \qquad (10.31)$$

SUGGESTIONS, SKILLS, AND STRATEGIES

You should know how to calculate the moment of inertia of a system of particles about a specified axis. The technique is straightforward, and consists of applying $I = \Sigma m_i r_i^2$, where m_i is the mass of the ith particle and r_i is the distance from the axis of rotation to the particle.

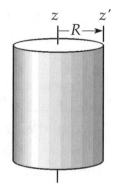

Once the moment of inertia about an axis through the center of mass I_{CM} is known, you can easily evaluate the moment of inertia about any axis parallel to the axis through the center of mass using the **parallel axis theorem** :

$$I = I_{CM} + Md^2$$

Figure 10.1

where d is the distance between the two axes.

For example, the moment of inertia of a solid cylinder about an axis through its center (the z axis in Figure 10.1) is given by $I_z = \frac{1}{2}MR^2$. Hence, the moment of inertia about the z' axis located a distance $d = R$ from the z axis is

$$I_{z'} = I_z + MR^2 = \tfrac{1}{2}MR^2 + MR^2 = \tfrac{3}{2}MR^2$$

This chapter includes a discussion of the application of Newton's laws in a special situation, namely, rigid bodies in static equilibrium. It is important that you understand and follow the procedures for analyzing such problems. The following skills must be mastered in this regard:

- Recognize all **external** forces acting on the body, and construct an accurate free-body diagram.

- Resolve the external forces into their rectangular components, and apply the first condition of equilibrium $\Sigma F_x = 0$ and $\Sigma F_y = 0$.

- Choose a convenient origin for calculating the net torque on the body. The choice of this origin is arbitrary. (The torque equation gives information which is not offered by applying $\Sigma \mathbf{F} = 0$.)

- Solve the set of simultaneous equations obtained from the two conditions of equilibrium.

The following procedure is recommended when analyzing a body in equilibrium under the action of several external forces:

- Make a sketch of the object under consideration.

- Draw a free-body diagram and label all external forces acting on the object. Try to guess the correct direction for each force. If you select an incorrect direction that leads to a negative sign in your solution for a force, do not be alarmed; this merely means that the direction of the force is the opposite of what you assumed.

- Resolve all forces into rectangular components, choosing a convenient coordinate system. Then apply the first condition for equilibrium, which balances forces. Remember to keep track of the signs of the various force components.

- Choose a convenient axis for calculating the net torque on the object. Remember that the choice of the origin for the torque equation is **arbitrary;** therefore, choose an origin that will simplify your calculation as much as possible. Becoming adept at this is a matter of practice.

- The first and second conditions of equilibrium give a set of linear equations with several unknowns. All that is left is to solve the simultaneous equations for the unknowns in terms of the known quantities.

The operation of the vector or cross product is used for the first time in this chapter. (Recall that the angular momentum $\mathbf{L}$ of a particle is defined as $\mathbf{L} = \mathbf{r} \times \mathbf{p}$, while torque is defined by the expression $\tau = \mathbf{r} \times \mathbf{F}$.) Let us briefly review the cross-product operation and some of its properties.

If $\mathbf{A}$ and $\mathbf{B}$ are any two vectors, their cross product, written as $\mathbf{A} \times \mathbf{B}$, is another vector $\mathbf{C}$. That is,

$$\mathbf{C} = \mathbf{A} \times \mathbf{B}$$

where the magnitude of $\mathbf{C}$ is given by

$$C = |\mathbf{C}| = AB\sin\theta$$

and θ is the angle between $\mathbf{A}$ and $\mathbf{B}$ as in Figure 10.2.

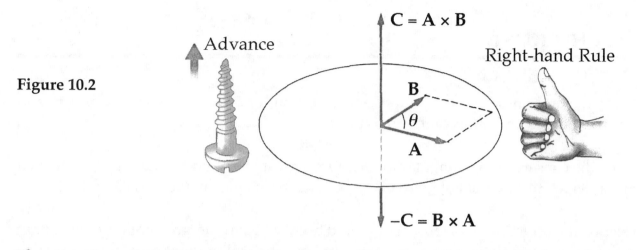

Figure 10.2

The direction of **C** is perpendicular to the plane formed by **A** and **B**, and its sense is determined by the right-hand rule. You should practice this rule for various choices of vector pairs. Note that **B** × **A** is directed **opposite** to **A** × **B**. That is, **A** × **B** = −**B** × **A**. This follows from the right-hand rule. You should not confuse the cross product of two vectors, which is a vector quantity, with the dot product of two vectors, which is a scalar quantity. (Recall that the dot product is defined as **A·B** = $AB \cos \theta$.)

Note that the cross product of any vector with itself is zero. That is, **A** × **A** = 0 since $\theta = 0°$, and $\sin(0) = 0$.

Very often, vectors will be expressed in unit vector form, and it is convenient to make use of the multiplication table for unit vectors. Note that **i**, **j**, and **k** represent a set of mutually orthogonal vectors as shown in Figure 10.3.

$$\mathbf{i} \times \mathbf{i} = \mathbf{j} \times \mathbf{j} = \mathbf{k} \times \mathbf{k} = 0$$

$$\mathbf{i} \times \mathbf{j} = -\mathbf{j} \times \mathbf{i} = \mathbf{k}$$

$$\mathbf{j} \times \mathbf{k} = -\mathbf{k} \times \mathbf{j} = \mathbf{i}$$

$$\mathbf{k} \times \mathbf{i} = -\mathbf{i} \times \mathbf{k} = \mathbf{j}$$

Figure 10.3

For example, if **A** = 3.0**i** + 5.0**j** and **B** = 4.0**j**, then we can take the cross product:

A × **B** = (3.0**i** + 5.0**j**) × (4.0**j**) = 3.0**i** × 4.0**j** + 5.0**j** × 4.0**j** = 12.0**k**

REVIEW CHECKLIST

▷ Define the cross product (magnitude and direction) of any two vectors, **A** and **B**, and state the various properties of the cross product.

▷ Apply the conservation of angular momentum principle to a body rotating about a fixed axis, in which the moment of inertia changes due to a change in the mass distribution.

▷ Describe the center of mass motion of a rigid body which undergoes both rotation about some axis and translation in space. Note that for pure rolling motion of an object such as a sphere or cylinder, the total kinetic energy can be expressed as the sum of a rotational kinetic energy about the center of mass plus the translational energy of the center of mass.

▷ Quantitatively, the angular displacement, speed, and acceleration for a rigid body system in rotational motion are related to the distance traveled, tangential speed, and tangential acceleration. The linear quantity is calculated by multiplying the angular quantity by the radius arm for an object or point in that system.

▷ If a body rotates about a fixed axis, every particle on the body has the same angular speed and angular acceleration. For this reason, rotational motion can be simply described using these quantities. The formulas which describe angular motion are analogous to the corresponding set of formulas pertaining to linear motion.

▷ Calculate the moment of inertia I of a system of particles or a rigid body about a specific axis. Note that the value of I depends on (a) the mass distribution and (b) the axis about which the rotation occurs. The parallel-axis theorem is useful for calculating I about an axis parallel to one that goes through the center of mass.

▷ Understand the concept of torque associated with a force, noting that the torque associated with a force has a magnitude equal to the force times the moment arm. Furthermore, note that the value of the torque depends on the origin about which it is evaluated.

▷ Recognize that the work-kinetic energy theorem can be applied to a rotating rigid body. That is, the net work done on a rigid body rotating about a fixed axis equals the change in its rotational kinetic energy.

▷ Describe the two necessary conditions of equilibrium for a rigid body.

▷ Analyze problems of rigid bodies in static equilibrium using the procedures presented in Section 10.6 of the text.

ANSWERS TO SELECTED CONCEPTUAL QUESTIONS

3. What is the magnitude of the angular velocity ω of the second hand of a clock? What is the direction of ω as you view a clock hanging vertically? What is the magnitude of the angular acceleration α of the second hand?

Answer   We use the rigid body under constant angular speed model.

The second hand of a clock turns at one revolution per minute, so

$$\omega = \frac{2\pi \text{ rad}}{60 \text{ s}} = 0.105 \text{ rad} / \text{s}$$

The motion is clockwise, so the direction of the vector angular velocity is away from you. It turns steadily, so ω is constant, and α is zero.

4. If you see an object rotating, does a net torque necessarily act on it?

Answer   An object rotates with constant angular momentum when zero total torque acts on it. For example, consider the Earth; it rotates at a constant rate of once per day, but there is no net torque acting on it.

10. A ladder rests inclined against a wall. Would you feel safer climbing up the ladder if you were told that the floor is frictionless but the wall is rough, or that the wall is frictionless but the floor is rough? Justify your answer.

Answer

The picture shows the forces on the ladder if both the wall and floor exert friction. If the floor is perfectly smooth, it can exert no frictional force to the right, to counterbalance the wall's normal force. Therefore a ladder on a smooth floor cannot stand in equilibrium. On the other hand, a smooth wall can still exert a normal force to hold the ladder in equilibrium against horizontal motion. The counterclockwise torque of this force prevents rotation about the foot of the ladder. So you should choose a rough floor.

13. Why do tightrope walkers carry a long pole to help balance themselves?

Answer

The long pole increases the tightrope walker's moment of inertia, and therefore decreases his angular acceleration, under any given torque. That gives him more time to respond, and in essence, aids his reflexes.

SOLUTIONS TO SELECTED END-OF-CHAPTER PROBLEMS

1. A motor rotating a grinding wheel at 100 rev/min is switched off. Assuming constant negative angular acceleration of magnitude 2.00 rad/s², (a) how long does it take the wheel to stop? (b) Through how many radians does it turn while it is slowing down?

Solution We use the rigid body under constant angular acceleration model.

We are given $\alpha = -2.00$ rad/s²

$\omega_f = 0$ $\qquad\qquad\qquad \omega_i = 100 \dfrac{\text{rev}}{\text{min}}\left(2\pi \dfrac{\text{rad}}{\text{rev}}\right)\left(\dfrac{1 \text{ min}}{60.0 \text{ s}}\right) = 10.47$ rad/s

(a) $\omega_f = \omega_i + \alpha t$ $\qquad t = \dfrac{\omega_f - \omega_i}{\alpha} = \dfrac{0 - (10.5 \text{ rad/s})}{-2.00 \text{ rad/s}^2} = 5.24$ s $\qquad\qquad \lozenge$

(b) $\omega_f^2 - \omega_i^2 = 2\alpha\left(\theta_f - \theta_i\right)$ $\qquad \theta_f - \theta_i = \dfrac{\omega_f^2 - \omega_i^2}{2\alpha} = \dfrac{0 - (10.5 \text{ rad/s})^2}{2\left(-2.00 \text{ rad/s}^2\right)} = 27.4$ rad $\quad \lozenge$

Note also in part (b) that since a constant acceleration is acting for time t,

$\theta_f - \theta_i = \bar{\omega} t = \left(\dfrac{10.5 + 0 \text{ rad/s}}{2}\right)(5.24 \text{ s}) = 27.4$ rad $\qquad\qquad\qquad\qquad \lozenge$

---

9. A disk 8.00 cm in radius rotates about its central axis at a constant rate of 1200 rev/min. Determine (a) its angular speed, (b) the tangential speed at a point 3.00 cm from its center, (c) the radial acceleration of a point on the rim, and (d) the total distance a point on the rim moves in 2.00 s.

Solution (a) $\omega = 2\pi f = (2\pi \text{ rad/rev})\left(\dfrac{1200 \text{ rev/min}}{60 \text{ s/min}}\right) = 125.7$ rad/s $= 126$ rad/s $\qquad \lozenge$

(b) $v = \omega R = (125.7 \text{ rad/s})(0.0300 \text{ m}) = 3.77$ m/s $\qquad\qquad\qquad\qquad\qquad \lozenge$

(c) $a_c = \omega^2 R = (125.7 \text{ rad/s})^2 (0.0800 \text{ m}) = 1.26 \times 10^3$ m/s² $\qquad\qquad\qquad \lozenge$

(d) $s = R\theta = R\omega t = \left(8.00 \times 10^{-2} \text{ m}\right)(125.7 \text{ rad/s})(2.00 \text{ s}) = 20.1$ m $\qquad\qquad \lozenge$

---

15. This problem describes one experimental method of determining the moment of inertia of an irregularly shaped object such as the payload for a satellite. Figure P10.15 shows a counterweight of mass m suspended by a cord that is wound around a spool of radius r. The spool forms part of a turntable that supports the object and is free to rotate without friction. When the counterweight is released from rest, it descends through a distance h, acquiring a speed v. Show that the moment of inertia I of the rotating object (including the turntable) is

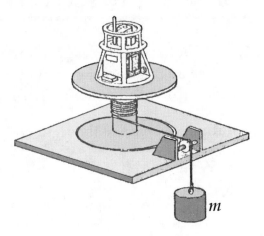

Figure P10.15

$$mr^2\left(2gh\,/\,v^2-1\right)$$

Solution

If the friction is negligible, then the energy of the counterweight-payload-turntable-Earth system is conserved as the counterweight unwinds. Each point on the cord moves at a linear speed of $v = \omega r$, where r is the radius of the spool. The energy conservation equation gives us:

$$\left(K_1 + K_2 + U_g\right)_i + W_{\text{other}} = \left(K_1 + K_2 + U_g\right)_f$$

Solving, we have

$$0 + 0 + mgh + 0 + 0 = \tfrac{1}{2}mv^2 + \tfrac{1}{2}I\omega^2 + 0 + 0$$

$$mgh = \tfrac{1}{2}mv^2 + \tfrac{1}{2}\frac{Iv^2}{r^2}$$

$$2mgh - mv^2 = I\frac{v^2}{r^2}$$

and finally,

$$I = mr^2\left(\frac{2gh}{v^2} - 1\right) \qquad \diamond$$

17. Find the net torque on the wheel in Figure P10.17 about the axle through O if $a = 10.0$ cm and $b = 25.0$ cm.

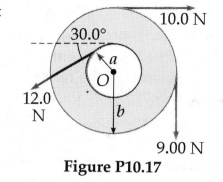

Figure P10.17

Solution $\sum \tau = \sum Fd$: $\tau_{net} = (12.0 \text{ N})(0.100 \text{ m})$
$-(10.0 \text{ N})(0.250 \text{ m})$
$-(9.00 \text{ N})(0.250 \text{ m})$

$\tau_{net} = -3.55 \text{ N} \cdot \text{m}$ ◊

This is $3.55 \text{ N} \cdot \text{m}$ into the plane of the page. ◊

Note that the 30° angle is not required for the solution. Note also that the 10.0-N and 9.00-N forces both produce clockwise, negative torques.

---

23. A uniform beam of mass m_b and length ℓ supports blocks of masses m_1 and m_2 at two positions, as in Figure P10.23. The beam rests on two knife edges. For what value of x will the beam be balanced at P such that the normal force at O is zero?

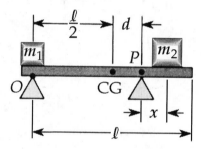

Figure P10.23

Solution We use the rigid body in equilibrium model.

Refer to the free-body diagram, and take torques about point P.

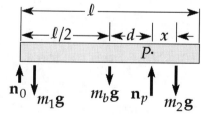

$$\sum \tau_P = -n_0 \left[\frac{\ell}{2} + d \right] + m_1 g \left[\frac{\ell}{2} + d \right] + m_b g d - m_2 g x = 0$$

We want to find x for which $n_0 = 0$. Let $n_0 = 0$ and solve for x.

$$m_1 \left[\frac{\ell}{2} + d \right] + m_b d - m_2 x = 0$$

so $$x = \frac{m_1}{m_2} \left[\frac{\ell}{2} + d \right] + \frac{m_b}{m_2} d$$ ◊

27. A uniform sign of weight F_g and width $2L$ hangs from a light, horizontal beam hinged at the wall and supported by a cable (Fig. P10.27). Determine (a) the tension in the cable and (b) the components of the reaction force exerted by the wall on the beam in terms of F_g, d, L, and θ.

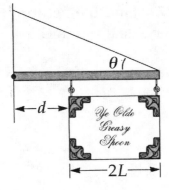

Solution

Choose the beam for analysis, and draw a free-body diagram as shown. We know that the direction of the force from the cable at the right end is along the cable, at an angle of θ above the horizontal.

Figure P10.27

Taking torques about the left end,

$$\Sigma F_x = 0: \qquad +R_x - T\cos\theta = 0$$

$$\Sigma F_y = 0: \qquad +R_y - F_g + T\sin\theta = 0$$

$$\Sigma \tau = 0: \qquad R_y(0) + R_x(0) - F_g(d+L) + (0)(T\cos\theta) + (d+2L)(T\sin\theta) = 0$$

(a) The torque equation gives

$$T = \frac{F_g(d+L)}{(d+2L)\sin\theta} \qquad \Diamond$$

(b) Now from the force equations,

$$R_x = \frac{F_g(d+L)}{(d+2L)\tan\theta}$$

and

$$R_y = F_g - \frac{F_g(d+L)}{d+2L} = \frac{F_g L}{d+2L} \qquad \Diamond$$

33. Two blocks, as shown in Figure P10.33, are connected by a string of negligible mass passing over a pulley of radius 0.250 m and moment of inertia I. The block on the frictionless incline is moving up with a constant acceleration of 2.00 m/s². (a) Determine T_1 and T_2, the tensions in the two parts of the string. (b) Find the moment of inertia of the pulley.

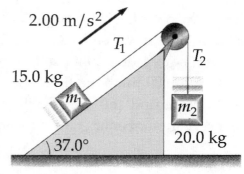

Figure P10.33

Solution

(a) The 15.0-kg block weighs
$$F_g = mg = (15.0 \text{ kg})(9.80 \text{ m/s}^2) = 147 \text{ N}$$

We assume the surface of the incline to be frictionless. Taking the x axis as directed up the incline, $\Sigma F_x = ma_x$ yields:

$$(-147 \text{ N})\sin 37° + T_1 = (15.0 \text{ kg})(2.00 \text{ m/s}^2)$$

$$T_1 = 118 \text{ N} \qquad \qquad \diamond$$

For the 20.0-kg block, we have $\Sigma F_y = ma_y$,   or

$$T_2 - (20.0 \text{ kg})(9.80 \text{ m/s}^2) = (20.0 \text{ kg})(-2.00 \text{ m/s}^2)$$

So, $T_2 = 156 \text{ N} \qquad \qquad \diamond$

(b) Now for the pulley, we use the rigid body under a net torque model.

$$\alpha = \frac{a}{r} = \frac{-2.00 \text{ m/s}^2}{0.250 \text{ m}} = -8.00 \text{ rad/s}^2, \quad \text{taking positive to be counterclockwise.}$$

$$\Sigma \tau = I\alpha, \quad \text{or} \quad (+118 \text{ N})(0.250 \text{ m}) - (156 \text{ N})(0.250 \text{ m}) = I(-8.00 \text{ rad/s}^2)$$

$$I = \frac{9.38 \text{ N}\cdot\text{m}}{8.00 \text{ rad/s}^2} = 1.17 \text{ kg}\cdot\text{m}^2 \qquad \qquad \diamond$$

35. An object with weight 50.0 N is attached to the free end of a light string wrapped around a reel with a radius of 0.250 m and a mass of 3.00 kg. The reel is a solid disk, free to rotate in a vertical plane about the horizontal axis passing through its center. The object is released 6.00 m above the floor. (a) Determine the tension in the string, the acceleration of the object, and the speed with which the object hits the floor. (b) Using the principle of conservation of energy, find the speed with which the object hits the floor.

Solution   Since the rotational inertia of the reel will slow the fall of the weight, we should expect the downward acceleration to be less than g. If the reel did not rotate, the tension in the string would be equal to the weight of the object; and if the reel disappeared, the tension would be zero. Therefore, $T < mg$ for the given problem. With similar reasoning, the final speed must be less than if the weight were to fall freely:

$$v_f < \sqrt{2g\Delta y} \cong 11 \text{ m / s}$$

We can find the acceleration and tension using the rotational form of Newton's second law. The final speed can be found using the particle under constant acceleration model, and also from conservation of energy. Free-body diagrams will greatly assist in analyzing the forces.

(a)   Use $\Sigma\tau = I\alpha$ to find T and a.

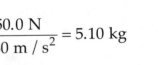

First find I for the reel, which we assume to be a uniform disk.

$$I = \tfrac{1}{2}MR^2 = \tfrac{1}{2}3.00 \text{ kg}(0.250 \text{ m})^2 = 0.0938 \text{ kg} \cdot \text{m}^2$$

The forces on it are shown, including a normal force exerted by its axle. From the diagram, we can see that the tension is the only unbalanced force causing the wheel to rotate.

$\Sigma\tau = I\alpha$ becomes $n(0) + F_g(0) + T(0.250 \text{ m}) = \left(0.0938 \text{ kg} \cdot \text{m}^2\right)(a/0.250 \text{ m})$ (1)

where we have applied $a_t = r\alpha$ to the point of contact between string and pulley.

The falling object has mass $m = \dfrac{F_g}{g} = \dfrac{50.0 \text{ N}}{9.80 \text{ m / s}^2} = 5.10 \text{ kg}$

For this object, $\Sigma F_y = ma_y$: $+T - 50.0 \text{ N} = (5.10 \text{ kg})(-a)$ (2)

Note that since we have defined upwards to be positive, the minus sign shows that its acceleration is downward. We now have our two equations in the unknowns T and a for the two connected objects. Substituting T from equation (2) into equation (1), we have

$$\left[50.0\text{ N} - (5.10\text{ kg})a\right](0.250\text{ m}) = \left(0.0938\text{ kg}\cdot\text{m}^2\right)(a/0.250\text{ m})$$

$$12.5\text{ N}\cdot\text{m} - (1.28\text{ kg}\cdot\text{m})a = (0.375\text{ kg}\cdot\text{m})a \quad\text{or}\quad 12.5\text{ N}\cdot\text{m} = a(1.65\text{ kg}\cdot\text{m})$$

$$a = 7.57\text{ m}/\text{s}^2 \qquad \Diamond$$

and $$T = 50.0\text{ N} - 5.10\text{ kg}\left(7.57\text{ m}/\text{s}^2\right) = 11.4\text{ N} \qquad \Diamond$$

For the motion of the object, $$v_f^2 = v_i^2 + 2a\left(x_f - x_i\right) = 0^2 + 2\left(7.57\text{ m}/\text{s}^2\right)(6.00\text{ m})$$

$$v_f = 9.53\text{ m}/\text{s}\ (\text{down}) \qquad \Diamond$$

(b) The conservation of energy principle can take account of multiple objects more easily than Newton's second law.

$$\left(K_1 + K_2 + U_g\right)_i = \left(K_1 + K_2 + U_g\right)_f$$

$$0 + 0 + m_1 g y_{1i} + 0 = \tfrac{1}{2}m_1 v_{1f}^2 + \tfrac{1}{2}I_2\omega_{2f}^2 + 0 + 0$$

Now note that $\omega = v/r$ as the string unwinds from the reel. Making substitutions,

$$50.0\text{ N}(6.00\text{ m}) = \tfrac{1}{2}(5.10\text{ kg})v_f^2 + \tfrac{1}{2}\left(0.0938\text{ kg}\cdot\text{m}^2\right)\left(\frac{v_f}{0.250\text{ m}}\right)^2$$

$$300\text{ N}\cdot\text{m} = \tfrac{1}{2}(5.10\text{ kg})v_f^2 + \tfrac{1}{2}(1.50\text{ kg})v_f^2$$

$$v_f = \sqrt{\frac{2(300\text{ N}\cdot\text{m})}{6.60\text{ kg}}} = 9.53\text{ m}/\text{s} \qquad \Diamond$$

As we should expect, both methods give the same final speed for the falling object, but the energy method is simpler. The acceleration is less than g, and the tension is less than the object's weight as we predicted. Now that we understand the effect of the reel's moment of inertia, this problem solution could be applied to solve other real-world pulley systems with masses that should not be ignored.

39. The position vector of a particle of mass 2.00 kg is given as a function of time by $\mathbf{r} = (6.00\mathbf{i} + 5.00t\mathbf{j})$ m. Determine the angular momentum of the particle about the origin, as a function of time.

Solution The velocity of the particle is $\quad \mathbf{v} = \dfrac{d\mathbf{r}}{dt} = \dfrac{d}{dt}(6.00\mathbf{i} \text{ m} + 5.00t\mathbf{j} \text{ m}) = 5.00\mathbf{j} \text{ m / s}$

The angular momentum is $\quad \mathbf{L} = \mathbf{r} \times \mathbf{p} = m\mathbf{r} \times \mathbf{v} = (2.00 \text{ kg})(6.00\mathbf{i} \text{ m} + 5.00t\mathbf{j} \text{ m}) \times 5.00\mathbf{j} \text{ m / s}$

$$\mathbf{L} = \left(60.0 \text{ kg} \cdot \text{m}^2 / \text{s}\right)\mathbf{i} \times \mathbf{j} + \left(50.0t \text{ kg} \cdot \text{m}^2 / \text{s}\right)\mathbf{j} \times \mathbf{j}$$

$$\mathbf{L} = (60.0\mathbf{k}) \text{ kg} \cdot \text{m}^2/\text{s}, \text{ constant in time} \qquad \lozenge$$

---

49. A cylinder of mass 10.0 kg rolls without slipping on a horizontal surface. At the instant its center of mass has a speed of 10.0 m/s, determine (a) the translational kinetic energy of its center of mass, (b) the rotational energy about its center of mass, and (c) its total energy.

Solution (a) $\qquad K_{\text{trans}} = \tfrac{1}{2} m v_{\text{CM}}^2 = \tfrac{1}{2}(10.0 \text{ kg})(10.0 \text{ m/s})^2 = 500 \text{ J} \qquad \lozenge$

(b) Call the radius of the cylinder R. An observer at the center sees the rough surface and the circumference of the cylinder moving at 10.0 m/s, so the angular speed of the cylinder is:

$$\omega = \frac{v_{\text{CM}}}{R} = \frac{10.0 \text{ m/s}}{R}$$

The moment of inertia about an axis through the center of mass is:

$$I_{\text{CM}} = \tfrac{1}{2} m R^2$$

so $\qquad K_{\text{rot}} = \tfrac{1}{2} I_{\text{CM}} \omega^2 = \left(\tfrac{1}{2}\right)\left[\tfrac{1}{2}(10.0 \text{ kg})R^2\right]\left(\dfrac{10.0 \text{ m/s}}{R}\right)^2 = 250 \text{ J} \qquad \lozenge$

(c) We can now add up the total energy:

$$K_{\text{tot}} = 500 \text{ J} + 250 \text{ J} = 750 \text{ J} \qquad \lozenge$$

---

55. A 60.0-kg woman stands at the rim of a horizontal turntable having a moment of inertia of 500 kg·m² and a radius of 2.00 m. The turntable is initially at rest and is free to rotate about a frictionless, vertical axle through its center. The woman then starts walking around the rim clockwise (as viewed from above the system) at a constant speed of 1.50 m/s relative to the Earth. (a) In what direction and with what angular speed does the turntable rotate? (b) How much work does the woman do to set herself and the turntable into motion?

Solution   The table rotates in a direction opposite to that in which the woman walks. We use the angular momentum version of the isolated system model. There are no external torques acting on the woman-turntable system; therefore, from conservation of angular momentum, we have $L_f = L_i = 0$.

(a) Therefore, $L_f = I_w \omega_w + I_t \omega_t = 0$ and $\omega_t = -\dfrac{I_w}{I_t}\omega_w$

Solving, $\omega_t = -\left(\dfrac{m_w r^2}{I_t}\right)\left(\dfrac{v_w}{r}\right) = -\dfrac{(60.0 \text{ kg})(2.00 \text{ m})(1.50 \text{ m/s})}{500 \text{ kg}\cdot\text{m}^2}$

$\omega_t = -0.360 \text{ rad/s} = 0.360 \text{ rad/s CCW}$ ◊

(b) Work done = ΔK: $W = K_f - 0 = \tfrac{1}{2}m_{\text{woman}}v_{\text{woman}}^2 + \tfrac{1}{2}I_{\text{table}}\omega_{\text{table}}^2$

$W = \tfrac{1}{2}(60.0 \text{ kg})(1.50 \text{ m/s})^2 + \tfrac{1}{2}\left(500 \text{ kg}\cdot\text{m}^2\right)(0.360 \text{ rad/s})^2 = 99.9 \text{ J}$ ◊

Related Questions: (a) Why is the angular momentum of the woman-turntable system conserved? (b) Why is the mechanical energy of this system not conserved? (c) Is the linear momentum of this system conserved?

(a)   Because the axle exerts no torque on the woman-plus-turntable system; only torques from outside the system can change the total angular momentum.

(b)   The internal forces, of the woman pushing backward on the turntable and of the turntable pushing forward on the woman, both do positive work, converting chemical into kinetic energy.

(c)   No. If the woman starts walking north, she pushes south on the turntable. Its axle holds it still against linear motion by pushing north on it, and this outside force delivers northward linear momentum into the system.

57. A long, uniform rod of length L and mass M is pivoted about a horizontal, frictionless pin passing through one end. The rod is released from rest in a vertical position, as in Figure P10.57. At the instant the rod is horizontal, find (a) its angular speed, (b) the magnitude of its angular acceleration, (c) the x and y components of the acceleration of its center of mass, and (d) the components of the reaction force at the pivot.

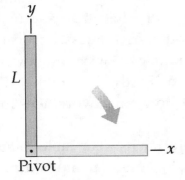

Figure P10.57

Solution

(a) Since only conservative forces are acting on the bar, use conservation of energy of the bar-Earth system:

$$\Delta K + \Delta U = 0$$

$$K_f - K_i + U_f - U_i = 0$$

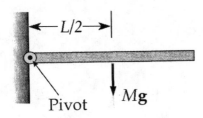

For evaluation of its gravitational energy, a rigid body can be modeled as a particle at its center of mass. Take the zero configuration for potential energy with the bar horizontal. Under these conditions $U_f = 0$ and $U_i = MgL/2$. Using the equation above,

$$\left(\tfrac{1}{2}I\omega_f^2 - 0\right) + \left(0 - \tfrac{1}{2}MgL\right)$$

and
$$\omega_f = \sqrt{MgL/I}$$

For a bar rotating about an axis through one end, $I = ML^2/3$.

Therefore,
$$\omega_f = \sqrt{(MgL)\big/\tfrac{1}{3}ML^2} = \sqrt{3g/L} \qquad \lozenge$$

(b) $\sum \tau = I\alpha:$ $\qquad Mg(L/2) = \left(\tfrac{1}{3}ML^2\right)\alpha,$ and $\qquad \alpha = 3g/2L$ $\qquad \lozenge$

(c) $a_x = a_c = r\omega_f^2 = \left(\dfrac{L}{2}\right)\left(\dfrac{3g}{L}\right) = \dfrac{3g}{2}$ ◊

Since this is **centripetal** acceleration, it is directed along the **negative** horizontal.

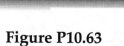

$a_y = a_t = r\alpha = \dfrac{L}{2}\alpha = \dfrac{3g}{4}$ ◊

(d) Using $\sum \mathbf{F} = m\mathbf{a}$, we have

$R_x = Ma_x = 3Mg/2$ in the **negative** direction ◊

$R_y - Mg = -Ma_y$ so $R_y = M\left(g - a_y\right) = M(g - 3g/4) = Mg/4$ ◊

---

63. A force acts on a uniform rectangular cabinet weighing 400 N, as in Figure P10.63. (a) If the cabinet slides with constant speed when $F = 200\,\text{N}$ and $h = 0.400\,\text{m}$, find the coefficient of kinetic friction and the position of the resultant normal force. (b) If $F = 300\,\text{N}$, find the value of h for which the cabinet just begins to tip.

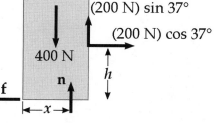

Figure P10.63

Solution

(a) Think of the normal force as acting at a distance x from the lower left corner. Moving with constant speed, the cabinet is in equilibrium:

$\Sigma F_x = 0$: $-f + (200\,\text{N})\cos(37.0°) = 0$

$\Sigma F_y = 0$: $-400\,\text{N} + n + (200\,\text{N})\sin(37.0°) = 0$

These equations tell us that $f = 160\,\text{N}$ and $n = 280\,\text{N}$, so $\mu_k = f/n = 0.571$ ◊

Take torques about the lower left corner; $\Sigma\tau = 0$ gives

$$-(400\text{ N})(30\text{ cm}) + nx + (200\text{ N})(60\text{ cm})\sin(37°) - (200\text{ N})(40\text{ cm})\cos(37°) = 0$$

Substituting $n = 280$ N (and converting cm to m, and back) gives

$$x = \frac{120\text{ N}\cdot\text{m} - 72.2\text{ N}\cdot\text{m} + 63.9\text{ N}\cdot\text{m}}{280\text{ N}} = 39.9\text{ cm} \qquad \Diamond$$

(b) When the cabinet is just about to tip, the normal force is located at the lower right corner, and $\Sigma\tau = 0$ is still true.

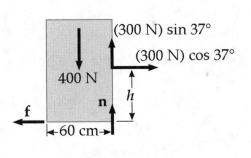

Because most of the forces are directed through the lower right corner, we choose to take torques about that point. This leaves only two forces to deal with.

$\Sigma\tau = 0$: $\qquad -(300\text{ N})(h)\cos(37.0°) + (400\text{ N})(30.0\text{ cm}) = 0$

Solving for h, $\qquad h = \dfrac{120\text{ N}\cdot\text{m}}{240\text{ N}} = 50.1\text{ cm} \qquad \Diamond$

Chapter 11

Gravity, Planetary Orbits, and the Hydrogen Atom

INTRODUCTION

We began our study of mechanics with translational motion and the forces that cause it. If we know the forces acting on a system and its initial conditions, we can predict its future. However, describing motion in this manner is often tedious and time-consuming. Fortunately, we can often follow the simpler approach of using conservation principles.

In this chapter we return to Newton's law of universal gravitation—one of the fundamental force laws in nature—and show how it, together with Newton's laws of motion, enables us to understand a variety of familiar orbital motions, such as the motions of planets and Earth satellites.

We conclude this chapter with a discussion of Niels Bohr's famous model of the hydrogen atom, which represents an interesting mixture of classical physics (Newton's laws of motion and the Coulomb interaction) and quantum physics (quantization of angular momentum). Although Bohr's theory contains ideas that are contrary to classical physics, his model successfully predicts the observed spectral lines of hydrogen.

NOTES FROM SELECTED CHAPTER SECTIONS

11.1 Newton's Law of Universal Gravitational Revisited

Newton's law of universal gravitation states that every particle in the universe attracts every other particle with a force that is directly proportional to the product of their masses and inversely proportional to the square of their distance of separation. The gravitational force always exists between two particles regardless of the medium which separates them.

The gravitational force exerted by a finite-size, spherically symmetric mass distribution on a particle outside the sphere is the same as if the entire mass of the sphere were concentrated at its center.

11.3 Kepler's Laws

Kepler deduced the following three empirical laws as they apply to our solar system.

- All planets move in elliptical orbits with the Sun at one of the focal points.
- The radius vector from the Sun to any planet sweeps out equal areas in equal times.
- The square of the orbital period of any planet is proportional to the cube of the semi-major axis for the elliptical orbit.

Kepler's second law is a consequence of the central nature of the gravitational force, and can be shown to follow from conservation of angular momentum. Kepler's third law follows from the inverse square nature of the case of a bound force.

11.4 Energy Considerations in Planetary and Satellite Motion

Both the total energy and the total angular momentum of a planet-Sun system are **constants of the motion**.

The **gravitational potential energy** for any pair of particles varies as $1/r$. The potential energy is negative since the force is one of attraction and the potential energy is taken to be zero when the distance of separation between the two particles is infinite. The absolute value of the potential energy is the **binding** energy of the system.

In the case of a bound gravitational system (closed orbits), the total energy must be negative. The kinetic energy in this case is positive and equal to one-half the magnitude of the potential energy.

The **escape speed** for an object projected from the Earth is independent of the mass of the object and it is independent of the direction of the initial speed (provided that the trajectory does not intersect Earth.) Also, when the initial speed is equal to the escape speed, the total energy is equal to zero.

11.5   Atomic Spectra and the Bohr Theory of Hydrogen

It is important to understand the behavior of the hydrogen atom as an atomic system for the following reasons:

- The quantum numbers used to characterize the allowed states of hydrogen can be used to describe the allowed states of more complex atoms. This enables us to understand the periodic table of the elements, which is one of the greatest triumphs of quantum mechanics.

- The hydrogen atom is an ideal system for performing precise tests of theory against experiment and for improving our overall understanding of atomic structure.

- Much of what is learned about the hydrogen atom with its single electron can be extended to such single-electron ions as He^+ and Li^{2+}, which are hydrogen-like in their atomic structure.

- The basic ideas about atomic structure must be well understood before we attempt to deal with the complexities of molecular structures and the electronic structure of solids.

The basic assumptions of the Bohr theory as it applies to the hydrogen atom are as follows:

- The electron moves in circular orbits about the proton under the influence of the Coulomb force of attraction. In this case, the Coulomb force causes the centripetal acceleration.

- Only certain electron orbits are stable. These are orbits in which the hydrogen atom does not emit energy in the form of radiation. Hence, the total energy of the atom remains constant.

- Radiation is emitted by the hydrogen atom when the electron "jumps" from a more energetic initial state to a lower state. The "jump" cannot be visualized or treated classically. In particular, the frequency, f, of the radiation emitted in the jump is related to the change in the atom's energy and is **independent of the frequency of the electron's orbital motion.** The frequency of the emitted radiation is found from

$$E_i - E_f = hf$$

where E_i is the energy of the initial state, E_f is the energy of the final state, h is Planck's constant, and $E_i > E_f$.

- The size of the allowed electron orbits is determined by a condition imposed on the electron's orbital angular momentum: The allowed orbits are those for which the electron's orbital angular momentum about the nucleus is an integral multiple of $\hbar = h/2\pi$.

$$mvr = n\hbar \quad \text{where} \quad n = 1, 2, 3, \ldots .$$

The lowest stationary state of an electron is called the ground state. The minimum energy required to ionize an atom (remove an electron in the ground state from the influence of the proton) is called the ionization energy.

According to the correspondence principle, quantum mechanics is in agreement with classical mechanics when the quantum numbers are very large.

EQUATIONS AND CONCEPTS

The **law of universal gravitation** states that any pair of particles **attract** each other with a force that is proportional to the product of their masses and inversely proportional to the square of their separation.

$$F_g = G\frac{m_1 m_2}{r^2} \qquad (11.1)$$

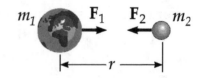

The constant G is called the **universal gravitational constant.**

$$G = 6.673 \times 10^{-11} \frac{\text{N} \cdot \text{m}^2}{\text{kg}^2} \qquad (11.2)$$

The gravitational force $\mathbf{F}$ can be expressed in vector form. The unit vector $\hat{\mathbf{r}}_{12}$ is directed from m_1 toward m_2, and $\mathbf{F}$ is the force on m_2 due to m_1. According to Newton's third law, $\mathbf{F}_{12} = -\mathbf{F}_{21}$.

$$\mathbf{F}_{12} = -\frac{G m_1 m_2}{r^2}\,\hat{\mathbf{r}}_{12} \qquad (11.3)$$

The constant of proportionality in Kepler's third law is independent of the mass of the planet. The value given is for orbits about the Sun. For orbits about the Earth, $K_{\text{E}} = 4\pi^2/GM_{\text{E}}$.

$$T^2 = \left(\frac{4\pi^2}{GM_S}\right)a^3 = K_S a^3 \qquad (11.5)$$

$$K_S = 2.97 \times 10^{-19}\ \text{s}^2/\text{m}^3$$

The total energy, E, of a two-body system, when the bodies are separated by a distance, r, is the sum of the kinetic energy of the orbiting object of mass m, and the potential energy of the system, where the object of mass M is assumed to be at rest in an inertial frame.

$$E = \frac{1}{2}mv^2 - G\frac{Mm}{r} \qquad (11.6)$$

As a body of mass m moves in a **circular orbit** around a massive body of mass M

$$E = -\frac{GMm}{2a} \qquad (11.8)$$

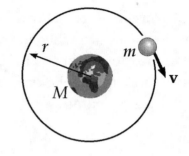

- Where $M \gg m$, and
- Defining the mass M to be at rest,

the **total energy** of the system is the sum of the kinetic energy of m and the potential energy of the system. When the two contributions are evaluated, one finds that the **total energy** E is negative and given by Eq. 11.8. This arises from the fact that the (positive) kinetic energy is equal to half the magnitude of the (negative) potential energy.

The **escape speed** is defined as the **minimum** speed a body must have, when projected from the Earth whose mass is M_E and radius is R_E, in order to escape the Earth's gravitational field (that is, to just reach $r = \infty$ with zero speed). Note that v_{esc} does not depend on the mass of the projected body. This equation can be applied to any object projected from any planet, by substituting M_E with M, and substituting R_E with R.

$$v_{esc} = \sqrt{\frac{2GM_E}{R_E}} \qquad (11.11)$$

The spectral lines which appear in the visible region of the hydrogen emission spectrum have wavelengths which can be calculated from an empirical equation. The resulting series of wavelength values is called the Balmer series.

$$\frac{1}{\lambda} = R_H\left(\frac{1}{2^2} - \frac{1}{n^2}\right) \qquad (11.14)$$

$$\text{where} \quad n = 3, 4, 5, \ldots$$

$$R_H = 1.0973732 \times 10^7 \text{ m}^{-1}$$

The electric potential energy of the hydrogen atom is given by $U_e = -k_e e^2/r$; the kinetic energy is $K = k_e e^2/2r$. The **total energy**, E, is the sum of the potential and kinetic energies. Note that the total energy is negative, which is indicative of a bound electron-proton system.

$$E = -\frac{k_e e^2}{2r} \tag{11.19}$$

An expression for the radii of the allowed orbits can be obtained by combining the equation for **quantization** of the orbital angular momentum of the electron with the equation for kinetic energy.

$$r_n = \frac{n^2 \hbar^2}{m_e k_e e^2} \tag{11.20}$$

where $n = 1, 2, 3, \ldots$

The quantization of the values of the electron orbit radii (as assumed in the Bohr theory) leads to a requirement of quantization of energy values for the electron.

$$E_n = -\frac{13.606}{n^2} \text{ eV} \tag{11.23}$$

where $n = 1, 2, 3, \ldots$

Radiation is emitted by an atom when the electron makes a transition from an initial state to a final state of different orbital radius. The frequency and wavelength of the emitted radiation can be calculated in terms of the initial and final quantum numbers.

$$\frac{1}{\lambda} = R_H \left(\frac{1}{n_f^2} - \frac{1}{n_i^2} \right) \tag{11.26}$$

and $\quad f = \frac{c}{\lambda}$

REVIEW CHECKLIST

▷ State Kepler's three laws of planetary motion and recognize that the laws are empirical in nature; that is, they are based upon astronomical data.

▷ Describe the nature of Newton's law of universal gravitation, and the method of deriving Kepler's third law ($T^2 \propto r^3$) from this law for circular orbits. Recognize that Kepler's second law is a consequence of conservation of angular momentum and the central nature of the gravitational force.

▷ Describe the total energy of a system of a planet or Earth satellite moving in a circular orbit about a large body located at the center of motion. Note that the total energy is negative, as it must be for any closed orbit.

▷ Understand the meaning of escape speed, and know how to obtain the expression for v_{esc} using the principle of conservation of energy.

▷ Describe the Bohr model of the hydrogen atom. Relate the basic assumptions and the observed spectral lines.

▷ Calculate the frequency and wavelength of the radiation that is emitted when an electron in hydrogen undergoes a transition between energy levels with different quantum numbers.

ANSWERS TO SELECTED CONCEPTUAL QUESTIONS

4. Explain why it takes more fuel for a spacecraft to travel from the Earth to the Moon than for the return trip. Estimate the difference.

Answer   The mass and radius of the Earth and Moon, and the distance between the two are $M_E = 5.98 \times 10^{24}$ kg, $R_E = 6.37 \times 10^6$ m, $M_M = 7.36 \times 10^{22}$ kg, $R_M = 1.75 \times 10^6$ m, and $d = 3.84 \times 10^8$ m. To travel between the Earth and the Moon, a distance d, a rocket engine must boost the spacecraft over the point of zero total gravitational field in between. Call x the distance of this point from Earth. To cancel, the Earth and Moon must here produce equal fields:

$$\frac{GM_E}{x^2} = \frac{GM_M}{(d-x)^2}$$

Isolating x, $\frac{M_E}{M_M}(d-x)^2 = x^2$ and $\sqrt{\frac{M_E}{M_M}}(d-x) = x$

Thus,

$$x = \frac{d\sqrt{M_E}}{\sqrt{M_M} + \sqrt{M_E}} = 3.46 \times 10^8 \text{ m}$$

and

$$(d - x) = \frac{d\sqrt{M_M}}{\sqrt{M_M} + \sqrt{M_E}} = 3.83 \times 10^7 \text{ m}$$

In general, we can ignore the gravitational pull of the far planet when we're close to the other; our results will still be approximately correct. Thus the approximate energy difference between point x and the Earth's surface, is:

$$\frac{\Delta E_{E \to x}}{m} = \frac{-GM_E}{x} - \frac{-GM_E}{R_E}$$

Similarly, flying from the moon:

$$\frac{\Delta E_{M \to x}}{m} = \frac{-GM_M}{d - x} - \frac{-GM_M}{R_M}$$

Taking the ratio of the energy terms,

$$\frac{\Delta E_{E \to x}}{\Delta E_{M \to x}} = \left(\frac{M_E}{x} - \frac{M_E}{R_E}\right) \Big/ \left(\frac{M_M}{d - x} - \frac{M_M}{R_M}\right) \cong 23.0$$

This would also be the minimum fuel ratio, if the spacecraft was driven by a method other than rocket exhaust. If rockets were used, Equation 8.42 would apply, and the total fuel ratio would be much larger.

□ □ □ □

7. Why don't we put a communications satellite in orbit around the 45th parallel? Wouldn't this be more useful in the United States and Canada than one in orbit around the Equator?

Answer While a satellite in orbit around the 45th parallel might be more useful, it isn't possible. The center of a satellite orbit must be the center of the Earth, since that is the force center for the gravitational force.

9. In his 1798 experiment, Cavendish was said to have "weighed the Earth." Explain this statement.

Answer The Earth creates a gravitational field at its surface according to $g = \dfrac{GM_E}{R_E^2}$. The factors g and R_E were known, so as soon as Cavendish measured G, he could compute the mass of the Earth.

□ □ □ □

SOLUTIONS TO SELECTED END-OF-CHAPTER PROBLEMS

5. The free-fall acceleration on the surface of the Moon is about one-sixth that on the surface of the Earth. If the radius of the Moon is about $0.250 R_E$, find the ratio of their average densities, ρ_{Moon}/ρ_{Earth}.

Solution The gravitational field at the surface of the Earth or Moon is given by $g = \dfrac{GM}{R^2}$.

The expression for density is
$$\rho = \frac{M}{V} = \frac{M}{\frac{4}{3}\pi R^3}$$

so
$$M = \frac{4}{3}\pi \rho R^3 \quad \text{and} \quad g = \frac{G\frac{4}{3}\pi \rho R^3}{R^2} = G\frac{4}{3}\pi \rho R$$

Noting that this equation applies to both the Moon and the Earth, and dividing the two equations,
$$\frac{g_M}{g_E} = \frac{G\frac{4}{3}\pi \rho_M R_M}{G\frac{4}{3}\pi \rho_E R_E} = \frac{\rho_M R_M}{\rho_E R_E}$$

Substituting, $\dfrac{1}{6} = \dfrac{\rho_M}{\rho_E}\left(\dfrac{1}{4}\right)$ and $\dfrac{\rho_M}{\rho_E} = \dfrac{4}{6} = \dfrac{2}{3}$ ◊

9. Compute the magnitude and direction of the gravitational field at a point P on the perpendicular bisector of two particles with equal masses separated by a distance $2a$ as shown in Figure P11.9.

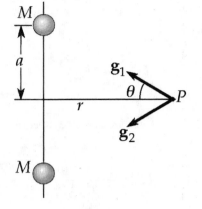

Solution We must add the vector fields created by each mass. In equation form, $\mathbf{g} = \mathbf{g}_1 + \mathbf{g}_2$ where

$$\mathbf{g}_1 = \frac{GM}{r^2 + a^2} \quad \text{to the left and upward at } \theta$$

and $\mathbf{g}_2 = \dfrac{GM}{r^2 + a^2}$ to the left and downward at θ

Figure P11.9
(modified)

Therefore,

$$\mathbf{g} = \frac{GM}{r^2 + a^2} \cos\theta(-\mathbf{i}) + \frac{GM}{r^2 + a^2}\sin\theta(\mathbf{j}) + \frac{GM}{r^2 + a^2}\cos\theta(-\mathbf{i}) + \frac{GM}{r^2 + a^2}\sin\theta(-\mathbf{j})$$

$$\mathbf{g} = \frac{2GM}{r^2 + a^2} \frac{r}{\sqrt{r^2 + a^2}}(-\mathbf{i}) + 0\mathbf{j} = \frac{-2GMr}{(r^2 + a^2)^{3/2}}\mathbf{i} \qquad \lozenge$$

---

11. After our Sun exhausts its nuclear fuel, its ultimate fate may be to collapse to a **white dwarf** state, in which it has approximately the same mass it has now but a radius equal to the radius of the Earth. Calculate (a) the average density of the white dwarf, (b) the free-fall acceleration due to gravity at its surface, and (c) the gravitational potential energy associated with a 1.00-kg object at its surface. (Take $U_g = 0$ at infinity.)

Solution

(a) $\rho = \dfrac{M_s}{V} = \dfrac{M_s}{\left(\frac{4}{3}\right)\pi R_E^{\ 3}} = \dfrac{1.99 \times 10^{30} \text{ kg}}{\left(\frac{4}{3}\right)\pi (6.37 \times 10^6 \text{ m})^3} = 1.84 \times 10^9 \text{ kg / m}^3 \qquad \lozenge$

(This is on the order of 1 million times the density of concrete!)

(b) For an object of mass m on its surface, $mg = GM_s m/R_E^2$. Thus,

$$g = \frac{GM_s}{R_E^2} = \frac{(6.67 \times 10^{-11} \text{ N} \cdot \text{m}^2/\text{kg}^2)(1.99 \times 10^{30} \text{ kg})}{(6.37 \times 10^6 \text{ m})^2} = 3.27 \times 10^6 \text{ m}/\text{s}^2 \qquad \lozenge$$

(This acceleration is on the order of 1 million times more than g_{Earth})

(c) $U_g = \dfrac{-GM_s m}{R_E} = \dfrac{(-6.67 \times 10^{-11} \text{ N} \cdot \text{m}^2/\text{kg}^2)(1.99 \times 10^{30} \text{ kg})(1 \text{ kg})}{(6.37 \times 10^6 \text{ m})} = -2.08 \times 10^{13} \text{ J} \qquad \lozenge$

(Such a large potential energy could yield a big gain in kinetic energy with even small changes in height. For example, dropping the 1.00-kg object from 1.00 m would result in a final velocity of 2 560 m/s.)

---

13. Plaskett's binary system consists of two stars that revolve in a circular orbit about a center of mass midway between them. This means that the masses of the two stars are equal (Figure P11.13). If the orbital velocity of each star is 220 km/s and the orbital period of each is 14.4 days, find the mass M of each star. (For comparison, the mass of our Sun is 1.99×10^{30} kg.)

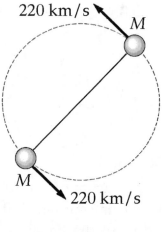

220 km/s

M

M

220 km/s

Figure P11.13

Solution   From the given data, it is difficult to estimate a reasonable answer to this problem without actually working through the details to actually solve it. A reasonable guess might be that each star has a mass larger than our Sun since fourteen days is short compared to the orbital periods of all the Sun's planets.

The only force acting on the two stars is the central gravitational force of attraction which results in a centripetal acceleration. When we solve Newton's second law, we can find the unknown mass in terms of the variables given in the problem.

Applying Newton's 2nd Law, $\Sigma F = ma$ yields $F_g = ma_c$ for each star:

$$\frac{GMM}{(2r)^2} = \frac{Mv^2}{r} \qquad \text{so} \qquad M = \frac{4v^2 r}{G}$$

We can write r in terms of the period, T, by considering the time and distance of one complete cycle. The distance traveled in one orbit is the circumference of the stars' common orbit, so $2\pi r = vT$. Therefore

$$M = \frac{4v^2r}{G} = \left(\frac{4v^2}{G}\right)\left(\frac{vT}{2\pi}\right) = \frac{2v^3T}{\pi G}$$

$$M = \frac{2(220 \times 10^3 \text{ m/s})^3(14.4 \text{ d})(86400 \text{ s/d})}{\pi\left(6.67 \times 10^{-11} \text{ N} \cdot \text{m}^2 / \text{kg}^2\right)} = 1.26 \times 10^{32} \text{ kg} \qquad \lozenge$$

The mass of each star is about 63 solar masses, much more than our initial guess! A quick check in an astronomy book reveals that stars over 8 solar masses are considered to be **heavyweight** stars, and astronomers estimate that the maximum theoretical limit is about 100 solar masses before a star becomes unstable. So these two stars are exceptionally massive!

---

15. Io, a satellite of Jupiter, has an orbital period of 1.77 days and an orbital radius of 4.22×10^5 km. From these data, determine the mass of Jupiter.

Solution   We use the particle under a net force and the particle in uniform circular motion models.

The gravitational force of Jupiter on Io provides the centripetal acceleration of Io:

$$\Sigma F_{Io} = M_{Io}a$$

$$\frac{GM_JM_{Io}}{r^2} = \frac{M_{Io}v^2}{r} = \frac{M_{Io}}{r}\left(\frac{2\pi r}{T}\right)^2 = \frac{4\pi^2 r M_{Io}}{T^2}$$

Thus, $$M_J = \frac{4\pi^2 r^3}{GT^2} = \frac{4\pi^2(4.22 \times 10^8 \text{ m})^3}{(6.67 \times 10^{-11} \text{ N} \cdot \text{m}^2 / \text{kg}^2)(1.77 \text{ d})^2}\left(\frac{1 \text{ d}}{86\ 400 \text{ s}}\right)^2\left(\frac{\text{N} \cdot \text{s}^2}{\text{kg} \cdot \text{m}}\right)$$

and $$M_J = 1.90 \times 10^{27} \text{ kg} \qquad \lozenge$$

---

21. A "treetop satellite" (Fig. P11.21) moves in a circular orbit just above the surface of a planet, assumed to offer no air resistance. Show that its orbital speed v and the escape speed from the planet are related by the expression $v_{esc} = \sqrt{2}v$.

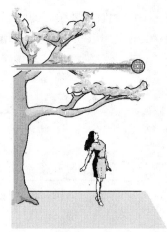

Solution   Call M the mass of the planet and R its radius. For the orbiting "treetop satellite," $\Sigma F = ma$ becomes

$$\frac{GMm}{R^2} = \frac{mv^2}{R} \qquad \text{or} \qquad v = \sqrt{\frac{GM}{R}}$$

If the object is launched with escape velocity, applying conservation of energy to the object-Earth system gives

$$\tfrac{1}{2}mv_{esc}^2 - \frac{GMm}{R} = 0 \qquad \text{or} \qquad v_{esc} = \sqrt{\frac{2GM}{R}}$$

Thus, $v_{esc} = \sqrt{2}v$ ◊

---

27. (a) What value of n_i is associated with the 94.96-nm spectral line in the Lyman hydrogen series? (b) Could this wavelength be associated with the Paschen or Balmer series?

Solution   Our equation is

$$\frac{1}{\lambda} = R_H\left(\frac{1}{n_f^2} - \frac{1}{n_i^2}\right)$$

where

$$R_H = 1.097 \times 10^7 \text{ m}^{-1}$$

and for the Lyman series,

$$n_f = 1, \text{ and } n_i = 2, 3, 4, \ldots$$

Substituting the given values,

$$\frac{1}{94.96 \times 10^{-9} \text{ m}} = \left(1.097 \times 10^{-9} \text{ m}^{-1}\right)\left(1 - \frac{1}{n_i^2}\right)$$

(a)   Solving for n_i, $n_i = 5$ ◊

(b)   By Figure 11.22, spectral lines in the Balmer and Paschen series all have much longer wavelengths, since much smaller energy losses put the atom into energy levels 2 or 3. ◊

---

29. A hydrogen atom is in its first excited state ($n = 2$). Using the Bohr theory of the atom, calculate (a) the radius of the orbit, (b) the linear momentum of the electron, (c) the angular momentum of the electron, (d) the kinetic energy, (e) the potential energy, and (f) the total energy.

Solution   We use Bohr's structural model of the hydrogen atom.

(a)   From Equations 11.20 and 11.21, $r_n = n^2 a_0$

$$r_2 = 4(0.0529 \text{ nm}) = 0.212 \text{ nm} \qquad \lozenge$$

Next, we can more easily do (c) before (b).

(c) $mvr = n\hbar = \dfrac{2(6.63 \times 10^{-34} \text{ J} \cdot \text{s})}{2\pi} = 2.11 \times 10^{-34} \text{ J} \cdot \text{s}$ $\lozenge$

(b) $mv = \dfrac{mvr}{r} = \dfrac{2.11 \times 10^{-34} \text{ J} \cdot \text{s}}{0.212 \times 10^{-9} \text{ m}} = 9.95 \times 10^{-25} \text{ kg} \cdot \text{m} / \text{s}$ $\lozenge$

(d) $K = \dfrac{1}{2} mv^2 = \dfrac{m^2 v^2}{2m} = \dfrac{p^2}{2m} = \dfrac{(9.95 \times 10^{-25} \text{ kg} \cdot \text{m} / \text{s})^2}{2(9.11 \times 10^{-31} \text{ kg})} = 5.44 \times 10^{-19} \text{ J}$

$$K = (5.44 \times 10^{-19} \text{ J}) \left(\dfrac{1 \text{ eV}}{1.60 \times 10^{-19} \text{ J}} \right) = 3.40 \text{ eV} \qquad \lozenge$$

(e) $U = \dfrac{-k_e e^2}{r} = \dfrac{(-8.99 \times 10^9 \text{ N} \cdot \text{m}^2 / \text{C}^2)(1.60 \times 10^{-19} \text{ C})^2}{(0.212 \times 10^{-9} \text{ m})} = -1.09 \times 10^{-18} \text{ J} = -6.80 \text{ eV}$ $\lozenge$

(f) $\qquad E = K + U = 3.40 \text{ eV} - 6.80 \text{ eV} = -3.40 \text{ eV}$ $\lozenge$

or $\qquad E = (-13.6 \text{ eV}) / n^2 = (-13.6 \text{ eV}) / 2^2 = -3.40 \text{ eV}$ $\lozenge$

---

37. The positron is the antiparticle to the electron. It has the same mass and a positive electric charge of the same magnitude as that of the electron. Positronium is a hydrogen-like atom consisting of a positron and an electron revolving around each other. Using the Bohr model, find the allowed distances between the two particles and the allowed energies of the system.

Solution

Since we are told that positronium is like hydrogen, we might expect the allowed radii and energy levels to be about the same as for hydrogen:

$$r = a_0 n^2 = \left(5.29 \times 10^{-11} \text{ m}\right) n^2 \qquad \text{and} \qquad E_n = (-13.6 \text{ eV}) / n^2$$

Similar to the textbook calculations for hydrogen, we can use the quantization of angular momentum of positronium to find the allowed radii and energy levels.

Let r represent the distance between the electron and the positron. The two move in a circle of radius $r/2$ around their center of mass with opposite velocities. The total angular momentum is quantized according to $\quad L_n = \frac{1}{2} mvr + \frac{1}{2} mvr = n\hbar \quad$ where $\quad n = 1, 2, 3, \ldots$

For each particle, $\Sigma F = ma$ expands to $\qquad \dfrac{k_e e^2}{r^2} = \dfrac{mv^2}{r/2}$

We can eliminate $v = \dfrac{n\hbar}{mr}$ to find $\qquad \dfrac{k_e e^2}{r} = \dfrac{2mn^2\hbar}{m^2 r^2}$

So the separation distances are $\qquad r = \dfrac{2n^2\hbar^2}{mk_e e^2} = 2a_0 n^2 = \left(1.06 \times 10^{-10} \text{ m}\right) n^2 \qquad \Diamond$

The orbital radii are $r/2 = a_0 n^2$, the same as for the electron in hydrogen.

The energy can be calculated from $\qquad E = K + U = \frac{1}{2} mv^2 + \frac{1}{2} mv^2 - \dfrac{k_e e^2}{r}$

Since $mv^2 = \dfrac{k_e e^2}{2r}$, $\qquad E = \dfrac{k_e e^2}{2r} - \dfrac{k_e e^2}{r} = -\dfrac{k_e e^2}{2r} = \dfrac{-k_e e^2}{4a_0 n^2} = -\dfrac{6.80 \text{ eV}}{n^2} \qquad \Diamond$

The electron moves in a circle of the same radius in our models of hydrogen and of positronium, but the allowed separation distances for positronium are twice as large as for hydrogen, and the steps between energy levels are half as big. One way to explain this is that in a hydrogen atom, the proton is much more massive than the electron, so the proton remains nearly stationary with essentially no kinetic energy. However, in positronium, the positron and electron have the same mass and therefore both have kinetic energy that separates them from each other and reduces the magnitude of their total energy compared with hydrogen.

43. In introductory physics laboratories, a typical Cavendish balance for measuring the gravitational constant G uses lead spheres of masses 1.50 kg and 15.0 g whose centers are separated by about 4.50 cm. Calculate the gravitational force between these spheres, treating each sphere as a point mass at the center of the sphere.

Solution $F = \dfrac{Gm_1m_2}{r^2} = \dfrac{\left(6.67 \times 10^{-11} \text{ N} \cdot \text{m}^2 / \text{kg}^2\right)(1.50 \text{ kg})(0.0150 \text{ kg})}{\left(4.50 \times 10^{-2} \text{ m}\right)^2}$

$F = 7.41 \times 10^{-10} \text{ N} = 741 \text{ pN}$ ◊

---

45. Two hypothetical planets of masses m_1 and m_2 and radii r_1 and r_2, respectively, are nearly at rest when they are an infinite distance apart. Because of their gravitational attraction, they head toward each other on a collision course. (a) When their center-to-center separation is d, find expressions for the speed of each planet and for their relative velocity. (b) Find the kinetic energy of each planet just before they collide if $m_1 = 2.00 \times 10^{24}$ kg, $m_2 = 8.00 \times 10^{24}$ kg, $r_1 = 3.00 \times 10^6$ m, and $r_2 = 5.00 \times 10^6$ m. (**Hint:** Both energy and momentum are conserved.)

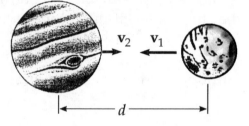

Solution   We use both the energy version and the momentum version of the isolated system model.

(a)   At infinite separation, $U = 0$; and at rest, $K = 0$. Since energy of the two-planet system is conserved, we have

$$0 = \tfrac{1}{2}m_1v_1^2 + \tfrac{1}{2}m_2v_2^2 - \dfrac{Gm_1m_2}{d} \tag{1}$$

The initial momentum of the system is zero and momentum is conserved. Therefore,

$$0 = m_1v_1 - m_2v_2 \tag{2}$$

Combine (1) and (2) to find $v_1 = m_2\sqrt{\dfrac{2G}{d(m_1 + m_2)}}$ and $v_2 = m_1\sqrt{\dfrac{2G}{d(m_1 + m_2)}}$

Relative velocity $v_r = v_1 - (-v_2) = \sqrt{\dfrac{2G(m_1 + m_2)}{d}}$ ◊

(b) Substitute the given numerical values into the equation found for v_1 and v_2 in part (a) to find $v_1 = 1.03 \times 10^4$ m/s   and $v_2 = 2.58 \times 10^3$ m/s.

Therefore, $K_1 = \frac{1}{2}m_1v_1{}^2 = 1.07 \times 10^{32}$ J   ◊

and $K_2 = \frac{1}{2}m_2v_2{}^2 = 2.67 \times 10^{31}$ J   ◊

---

51. Two stars of masses M and m, separated by a distance d, revolve in circular orbits about their center of mass (Fig. P11.51). Show that each star has a period given by

$$T^2 = \frac{4\pi^2}{G(M+m)}d^3$$

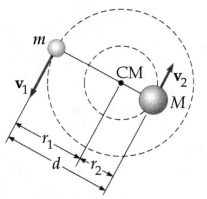

(**Hint:** Apply Newton's second law to each star, and note that the center-of-mass condition requires that $Mr_2 = mr_1$, where $r_1 + r_2 = d$.)

Solution   For the star of mass M and orbital radius r_2,

Figure P11.51

$\Sigma F = ma$ gives

$$\frac{GMm}{d^2} = \frac{Mv_2{}^2}{r_2} = \frac{M}{r_2}\left(\frac{2\pi r_2}{T}\right)^2$$

For the star of mass m, $\Sigma F = ma$ gives

$$\frac{GMm}{d^2} = \frac{mv_1{}^2}{r_1} = \frac{m}{r_1}\left(\frac{2\pi r_1}{T}\right)^2$$

Clearing fractions, we then obtain simultaneous equations:

$$GmT^2 = 4\pi^2 d^2 r_2 \qquad\qquad GMT^2 = 4\pi^2 d^2 r_1$$

Adding, we find $G(M+m)T^2 = 4\pi^2 d^2(r_1 + r_2) = 4\pi^2 d^3$

$$T^2 = \frac{4\pi^2 d^3}{G(M+m)}$$ ◊

In a visual binary star system T, d, r_1, and r_2 can sometimes be measured, so the mass of each component can be computed.

Chapter 12

Oscillatory Motion

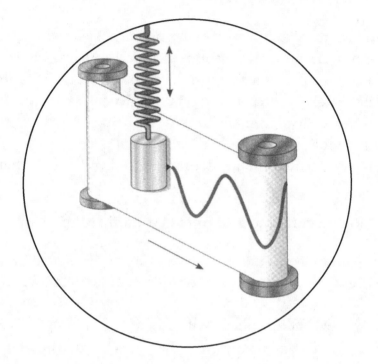

INTRODUCTION

A very special kind of motion occurs when the force on a body is proportional to the displacement of the body from equilibrium. If this force always acts toward the equilibrium position of the body, there is a repetitive back-and-forth motion about this position. Such motion is an example of what is called **periodic** or **oscillatory** motion.

You are most likely familiar with several examples of periodic motion, such as the oscillations of a mass on a spring, the motion of a pendulum, and the vibrations of a stringed musical instrument.

Most of the material in this chapter deals with **simple harmonic motion.** In this type of motion, an object oscillates between two spatial positions for an indefinite period of time with no loss in mechanical energy, and the position of the object is described as a sinusoidal function of time. In real mechanical systems, retarding (frictional) forces are always present and these forces are considered in an optional section at the end of the chapter.

NOTES FROM SELECTED CHAPTER SECTIONS

12.1 Motion of a Particle Attached to a Spring

Oscillatory motions are exhibited by many physical systems such as a mass attached to a spring, a pendulum, atoms in a solid, stringed musical instruments, and electrical circuits driven by a source of alternating current. An object exhibits **simple harmonic motion** if the net external force acting on it is a **linear restoring force**; the acceleration is proportional to the displacement from equilibrium and is in the opposite direction. Simple harmonic motion of a mechanical system corresponds to the oscillation of an object between two points for an indefinite period of time, with no loss in mechanical energy.

12.2 Mathematical Representation of Simple Harmonic Motion

The value of the phase constant ϕ depends on the initial position and initial velocity of the body. Figure 12.1 represents plots of the position, velocity, and acceleration versus time assuming that at $t = 0$, $x_i = A$ and $v_i = 0$. In this case, one finds that $\phi = 0$. Note that the velocity is 90° out of phase with the position. That is, v is zero when $|x|$ is a maximum, while $|v|$ is a maximum when x is zero. Furthermore, note that the acceleration is 180° out of phase with the position. That is, when x is a maximum and positive, a is a maximum, and negative. In other words, a is proportional to x, but in the **opposite** direction.

Figure 12.1

Representation of (a) the position, (b) the velocity, and (c) the acceleration as a function of time for an object moving with simple harmonic motion.

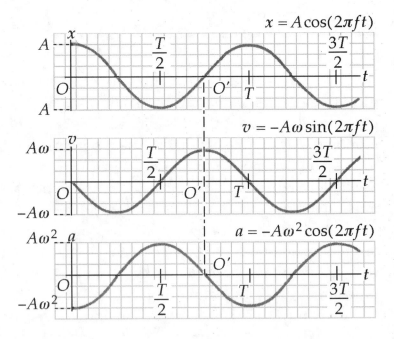

$$x = A\cos(2\pi f t)$$

$$v = -A\omega \sin(2\pi f t)$$

$$a = -A\omega^2 \cos(2\pi f t)$$

It is necessary to define a few terms relative to harmonic motion:

- The amplitude, A, is the **maximum distance that an object moves away from its equilibrium position.** In the absence of friction, an object will continue in simple harmonic motion. During each cycle, it will reach a maximum position on each side of the equilibrium position equal to the amplitude.

- The period, T, **is the time it takes the object to execute one complete cycle of the motion.**

- The frequency, f, **is the number of cycles or vibrations per second.**

A common system which undergoes simple harmonic motion is the block-spring system shown in Figure 12.2. The block is assumed to move on a horizontal, **frictionless** surface. The point $x = 0$ is the equilibrium position of the block; that is, the point where the block would reside if left undisturbed. In this position, there is no horizontal force on the block. When the block is displaced a distance x from its equilibrium position, the spring produces a linear restoring force given by Hooke's law, $F = -kx$, where k is the force constant of the spring, and has SI units of N/m. The minus sign means that F is to the **left** when the displacement x is positive, whereas F is to the **right** when x is negative. In other words, the direction of the force F is **always** towards the equilibrium position.

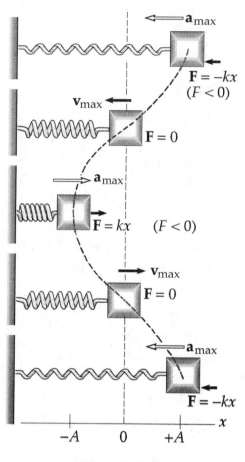

Figure 12.2

Oscillating motion of a block on the end of a spring.

12.3 Energy Considerations in Simple Harmonic Motion

You should study carefully the comparison between the motion of the mass-spring system and that of the simple pendulum. In particular, notice that when the position is a maximum, the energy of the system is entirely potential energy; whereas, when the position is zero, the energy is entirely kinetic energy. This is consistent with the fact that $v = 0$ when $|x| = A$, while $v = v_{max}$ when $x = 0$. For an arbitrary value of x, the energy is the sum of K and U.

12.4 The Simple Pendulum

A **simple pendulum** consists of an object of mass m attached to a light string of length L as shown in Figure 12.3. When the angular position θ is small during the entire motion (less than about 15°), the pendulum exhibits simple harmonic motion. In this case, the resultant force acting on the object equals the component of weight **tangent** to the circle, and has a magnitude $mg \sin \theta$. Since this force is always directed towards $\theta = 0$, it corresponds to a restoring force. For small θ, with θ measured in radians, we use the small angle approximation $\sin\theta \cong \theta$. In this approximation, the equation of motion reduces to Equation 12.23, $d^2\theta / dt^2 = -g\theta/L$.

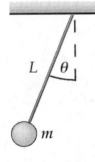

Figure 12.3

This equation is **identical** in form to Eq. 12.5 for the block-spring system, $d^2x/dt^2 = -\omega^2 x$, which has the solution $x = A\cos(\omega t + \phi)$. The corresponding solution to Equation 12.23 is $\theta = \theta_i \cos(\omega t + \phi)$, where ω is given by Equation 12.24. The period of motion is given by Equation 12.25. In other words, the period depends only on the length of the pendulum and the acceleration due to gravity. The period **does not** depend on mass, so we conclude that **all** simple pendula of equal length oscillate with the same frequency and period.

12.6   Damped Oscillations

Damped oscillations occur in realistic systems in which retarding forces such as friction are present. These forces will reduce the amplitudes of the oscillations with time, since mechanical energy is continually lost by the system. When the retarding force is assumed to be proportional to the velocity, but small compared to the restoring force, the system will still oscillate, but the amplitude will decrease exponentially with time.

It is possible to compensate for the energy lost in a damped oscillator by adding an additional driving force that does positive work on the system. This additional energy supplied to the system must at least equal the energy lost due to friction to maintain constant amplitude. The energy transferred to the system is a maximum when the driving force is in phase with the velocity of the system. The amplitude is a maximum when the frequency of the driving force matches the natural (resonance) frequency of the system.

EQUATIONS AND CONCEPTS

The force exerted by a spring on a block attached to the spring and displaced a distance x from the unstretched position is given by Hooke's law. The force constant, k, is always positive and has a value which corresponds to the relative stiffness of the spring. The negative sign means that the force exerted on the block is always directed opposite the position — the force is a restoring force, always directed toward the equilibrium position.

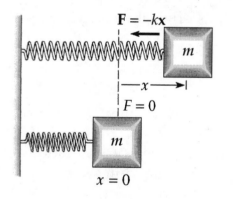

An object exhibits simple harmonic motion when the net force along the direction of motion is proportional to the displacement and in the opposite direction.

$$F_s = -kx \tag{12.1}$$

Applying **Newton's second law** to motion in the x direction gives $F = ma_x = -kx$. Since $a_x = d^2x/dt^2$, this is equivalent to Equation 12.5.

$$\frac{d^2x}{dt^2} = -\omega^2 x \tag{12.5}$$

The general solution to Equation 12.5 represents the time dependent position x, provided that $\omega^2 = k/m$. In this expression, A represents the **amplitude** of the motion, $\omega t + \phi$ is the **phase**, ω is the **angular frequency** (rad/s), and ϕ is the **phase constant**.

$$x(t) = A\cos(\omega t + \phi)$$

$$\text{with } \omega^2 = \frac{k}{m} \tag{12.6}$$

The **period** T equals the time it takes the block to complete **one** oscillation; that is, the time it takes the block to return to its original position with its original velocity for the first time.

$$T = \frac{2\pi}{\omega} = 2\pi\sqrt{\frac{m}{k}} \tag{12.13}$$

The **frequency** of the motion, f, numerically equals the inverse of the period and represents the number of oscillations per unit time. T is measured in seconds, while f is measured in s^{-1} or hertz (Hz).

$$f = \frac{1}{T} = \frac{1}{2\pi}\sqrt{\frac{k}{m}} \tag{12.14}$$

Taking the first derivative of x with respect to time gives the **velocity of the block as a function of time.**

$$v = \frac{dx}{dt} = -\omega A\sin(\omega t + \phi) \tag{12.15}$$

$$v_{max} = \omega A \tag{12.17}$$

The acceleration of the block as a function of time is equal to the time derivative of the velocity (or the second derivative of the displacement). Note that the **acceleration** (and hence the force) **is always proportional to and opposite the displacement.**

$$a = \frac{d^2x}{dt^2} = -\omega^2 A \cos(\omega t + \phi)$$ (12.16)

$$a_{max} = \omega^2 A$$ (12.18)

The kinetic energy of a simple harmonic oscillator is given by $\frac{1}{2}mv^2$, while the potential energy is equal to $\frac{1}{2}kx^2$. Using Equations 12.6 and 12.15, together with $\omega^2 = k/m$, gives the **total** energy E of the oscillator. Note that E remains constant since we have assumed there are no nonconservative forces acting on the system. The total energy of the simple harmonic oscillator is a constant of the motion and is proportional to the square of the amplitude.

$$E = \frac{1}{2}mv^2 + \frac{1}{2}kx^2$$

or

$$E = \frac{1}{2}kA^2$$ (12.21)

Energy conservation can be used to obtain an expression for velocity as a function of position.

$$v = \pm\sqrt{\frac{k}{m}\left(A^2 - x^2\right)}$$ (12.22)

or

$$v = \pm\omega\sqrt{\left(A^2 - x^2\right)}$$

The speed of an object in simple harmonic motion is a maximum at $x = 0$; the speed is zero when the object is at the points of maximum position $(x = \pm A)$.

The equation of motion for the simple pendulum assumes a small displacement so that $\sin\theta \cong \theta$.

$$\frac{d^2\theta}{dt^2} = -\frac{g}{L}\theta$$ (12.23)

The period and frequency of a simple pendulum depend only on the length of the supporting string and the value of the acceleration due to gravity.

$$\omega = \sqrt{\frac{g}{L}}$$ (12.24)

$$T = 2\pi\sqrt{\frac{L}{g}}$$ (12.25)

SUGGESTIONS, SKILLS, AND STRATEGIES

Most of this chapter deals with simple harmonic motion, and the properties of the position expression

$$x(t) = A\cos(\omega t + \phi)$$ (12.6)

In order to obtain the velocity $v(t)$ and acceleration $a(t)$ of the system, one must be familiar with the derivative operation as applied to trigonometric functions. In particular, note that

$$\frac{d}{dt}\cos(\omega t + \phi) = -\omega\sin(\omega t + \phi) \qquad \frac{d}{dt}\sin(\omega t + \phi) = \omega\cos(\omega t + \phi)$$

Using these results, and $x(t)$ from Equation 12.6, we see that

$$v(t) = \frac{dx(t)}{dt} = -A\omega\sin(\omega t + \phi)$$ (12.15)

and $$a(t) = \frac{dv(t)}{dt} = -A\omega^2\cos(\omega t + \phi)$$ (12.16)

By direct substitution, you should be able to show that Equation 12.6 represents a general solution to the equation of motion for the block-spring system (a second-order homogeneous differential equation) given by

$$\frac{d^2x}{dt^2} + \frac{k}{m}x = 0 \qquad \text{where} \qquad \omega = \sqrt{\frac{k}{m}}$$

In treating the motion of the simple pendulum, we made use of the small angle approximation $\sin\theta \cong \theta$. This approximation enables us to reduce the equation of motion to that of the simple harmonic oscillator. The small angle approximation for $\sin\theta$ follows from inspecting the series expansion for $\sin\theta$, where θ is in **radians**:

$$\sin\theta = \theta - \frac{\theta^3}{3!} + \frac{\theta^5}{5!} - \cdots$$

For small values of θ, the higher order terms in θ^3, θ^5 ... are **small** compared to θ, so it follows that $\sin\theta \cong \theta$. The difference between $\sin\theta$ and θ is less than 1% for $0 < \theta < 15°$ (where $15° \cong 0.26$ rad).

REVIEW CHECKLIST

▷ Describe the general characteristics of simple harmonic motion, and the significance of the various parameters which appear in the position function, $x = A\cos(\omega t + \phi)$.

▷ Start with the expression for the position for the simple harmonic oscillator, and obtain equations for the velocity and acceleration as functions of time.

▷ Understand the phase relations between position, velocity, and acceleration for simple harmonic motion, noting that acceleration is proportional to the position, but in the opposite direction.

▷ Describe and understand the conditions of simple harmonic motions executed by the block-spring system (where the frequency depends on k and m) and the simple pendulum (where the frequency depends on L and g).

ANSWERS TO SELECTED CONCEPTUAL QUESTIONS

4. If the position of a particle varies as $x = -A \cos \omega t$, what is the phase constant in Equation 12.6? At what position is the particle at $t = 0$?

Answer Equation 12.6 says $x = A \cos(\omega t + \phi)$. Since negating a cosine wave is equivalent to changing the phase by 180°, we can say that

$$\phi = 180° \left(\frac{\pi \, \text{rad}}{180°} \right) = \pi \, \text{rad}$$

At time $t = 0$, the particle is at a position of $x = -A$.

□ □ □ □

6. Determine whether the following quantities can have the same sign for a simple harmonic oscillator: (a) position and velocity, (b) velocity and acceleration, (c) position and acceleration.

Answer In a simple harmonic oscillator, the velocity follows the position by 1/4 of a cycle, and the acceleration follows the position by 1/2 of a cycle. Referring to Figure 12.1 of this study guide, it can be noted that there exist times (a) when both the position and the velocity are positive, and therefore in the same direction. (b) There also exist times when both the velocity and the acceleration are positive, and therefore in the same direction. On the other hand, (c) when the position is positive, the acceleration is always negative, and therefore position and acceleration always have opposite signs.

□ □ □ □

13. Is it possible to have damped oscillations when a system is at resonance? Explain.

Answer Yes. At resonance, the amplitude of a damped oscillator will remain constant. If the system were not damped, the amplitude would increase without limit at resonance.

□ □ □ □

SOLUTIONS TO SELECTED END-OF-CHAPTER PROBLEMS

3.   The position of a particle is given by the expression $x = (4.00 \text{ m}) \cos(3.00\pi t + \pi)$, where x is in meters and t is in seconds. Determine (a) the frequency and period of the motion, (b) the amplitude of the motion, (c) the phase constant, and (d) the position of the particle at $t = 0.250$ s.

Solution   We use the particle in simple harmonic motion model.

The particular position function $x = (4.00 \text{ m})\cos(3.00\pi t + \pi)$ and the general one, $x = A\cos(\omega t + \phi)$ have a specially powerful kind of equality called functional equality. They must give the same x value for all values of the variable t. This requires, then, that all parts be the same:

(a) $\omega = 3.00\pi = 2\pi f$ or $f = 1.50$ Hz; $T = \dfrac{1}{f} = 0.667$ s   ◊

(b) $A = 4.00$ m   ◊

(c) $\phi = \pi$ rad   ◊

(d) $x(t = 0.250 \text{ s}) = (4.00 \text{ m})\cos(1.75\pi \text{ rad}) = (4.00 \text{ m})\cos(5.50 \text{ rad})$.

   Note that this is **not** 5.50°. Instead,

   $x = (4.00 \text{ m})\cos(5.50 \text{ rad}) = (4.00 \text{ m})\cos 315° = 2.83$ m   ◊

---

7.   A particle moving along the x axis in simple harmonic motion starts from its equilibrium position, the origin, at $t = 0$ and moves to the right. If the amplitude of its motion is 2.00 cm and the frequency is 1.50 Hz, (a) show that the position of the particle is given by $x = (2.00 \text{ cm})\sin(3.00\pi t)$. Determine (b) the maximum speed and the earliest time $(t > 0)$ at which the particle has this speed, (c) the maximum acceleration and the earliest time $(t > 0)$ at which the particle has this acceleration to the right, and (d) the total distance traveled between $t = 0$ and $t = 1.00$ s.

Solution

(a)   At $t = 0$, $x = 0$ and v is positive (to the right). The sine function is zero and the cosine is positive at $\theta = 0$, so this situation corresponds to $x = A \sin \omega t$ and $v = v_i \cos \omega t$.

Since $f = 1.50$ Hz, $\omega = 2\pi f = 3\pi$

Also, $A = 2.00$ cm,   so that $x = (2.00 \text{ cm}) \sin(3\pi t)$ ◊

(b)   This is equivalent to writing $x = A \cos(\omega t + \phi)$

with $A = 2.00$ cm, $\omega = 3.00\pi/\text{s}$ and $\phi = -90° = -\dfrac{\pi}{2}$

Note also that $T = \dfrac{1}{f} = 0.667$ s

The velocity is $v = \dfrac{dx}{dt} = 2.00(3.00\pi)\cos(3.00\pi t) \text{ cm / s}$

The maximum speed is $v_{max} = v_i = A\omega = 2.00(3.00\pi) \text{ cm/s} = 6.00\pi \text{ cm/s}$ ◊

The particle has this speed at $t = 0$, when $\cos(3.00\pi t) = +1$,

and next at $t = \dfrac{T}{2} = 0.333$ s, when $\cos\left[\left(3.00\pi \text{ s}^{-1}\right)(0.333 \text{ s})\right] = -1$ ◊

(c)   Again, $a = \dfrac{dv}{dt} = (-2.00 \text{ cm})(3.00\pi \text{ s}^{-1})^2 \sin(3.00\pi t)$

Its maximum value is $a_{max} = A\omega^2 = (2.00 \text{ cm})(3.00\pi \text{ s}^{-1})^2 = 178 \text{ cm / s}^2$ ◊

The acceleration has this positive value for the first time at $t = \dfrac{3T}{4} = 0.500$ s   ◊

when $a = -(2.00 \text{ cm})(3.00\pi \text{ s}^{-1})^2 \sin\left[\left(3.00\pi \text{ s}^{-1}\right)(0.500 \text{ s})\right] = 178 \text{ cm / s}^2$

(d)   Since $T = \dfrac{2}{3}$ s and $A = 2.00$ cm, the particle will travel 8.00 cm in this time.

Hence, in $(1.00 \text{ s}) = \left(\dfrac{3T}{2}\right)$ the particle will travel $\dfrac{3}{2}(8.00 \text{ cm}) = 12.0$ cm   ◊

9. A 7.00-kg object is hung from the bottom end of a vertical spring fastened to an overhead beam. The object is set into vertical oscillations with a period of 2.60 s. Find the force constant of the spring.

Solution An object hanging from a vertical spring moves with simple harmonic motion just like an object moving without friction attached to a horizontal spring.

We are given the period; by Equation 12.13,
$$\frac{2\pi}{\omega} = T = 2.60 \text{ s}$$

Solving for the angular frequency,
$$\omega = \frac{2\pi \text{ rad}}{2.60 \text{ s}} = 2.42 \text{ rad / s}$$

However, $\omega = \sqrt{\dfrac{k}{m}}$, so
$$k = \omega^2 m = (2.41 \text{ rad / s})^2 (7.00 \text{ kg})$$

Thus we solve for the force constant:
$$k = 40.9 \text{ kg / s}^2 = 40.9 \text{ N / m} \qquad \Diamond$$

---

11. A 0.500-kg object attached to a spring of force constant 8.00 N/m vibrates in simple harmonic motion with an amplitude of 10.0 cm. Calculate (a) the maximum value of its speed and acceleration, (b) the speed and acceleration when the object is 6.00 cm from the equilibrium position, and (c) the time it takes the object to move from $x = 0$ to $x = 8.00$ cm.

Solution $\omega = \sqrt{\dfrac{k}{m}} = \sqrt{\dfrac{8.00 \text{ N / m}}{0.500 \text{ kg}}} = 4.00 \text{ s}^{-1}$

Therefore, position is given by $x = (10.0 \text{ cm})\sin[(4.00 \text{ s}^{-1})t]$. From this we find that

(a) $v = \dfrac{dx}{dt} = (40.0 \text{ cm / s})\cos(4.00t)$ $v_{max} = 40.0 \text{ cm/s}$ $\qquad \Diamond$

$a = \dfrac{dv}{dt} = -(160 \text{ cm / s}^2)\sin(4.00t)$ $a_{max} = 160 \text{ cm/s}^2$ $\qquad \Diamond$

(b) $t = \frac{1}{4}\sin^{-1}\left(\dfrac{x}{10.0 \text{ cm}}\right)$

When $x = 6.00$ cm, $t = 0.161$ s and we find that

$$v = (40.0 \text{ cm / s})\cos\left[(4.00 \text{ s}^{-1})(0.161 \text{ s})\right] = 32.0 \text{ cm / s} \qquad \lozenge$$

$$a = -(160 \text{ cm / s}^2)\sin\left[(4.00 \text{ s}^{-1})(0.161 \text{ s})\right] = -96.0 \text{ cm / s}^2 \qquad \lozenge$$

(c) Using $t = \frac{1}{4}\sin^{-1}\left(\dfrac{x}{10.0 \text{ cm}}\right)$

When $x = 0$, $t = 0$ and when $x = 8.00$ cm, $t = 0.232$ s. Therefore $\Delta t = 0.232$ s $\qquad \lozenge$

---

15. An automobile having a mass of 1000 kg is driven into a brick wall in a safety test. The bumper behaves like a spring of constant 5.00×10^6 N/m, and compresses 3.16 cm as the car is brought to rest. What was the speed of the car before impact, assuming that the mechanical energy of the car remains constant during impact with the wall?

Solution   We simplify by assuming that the car does no work on the wall or on the surrounding air. That is, we model the wall as perfectly solid and we ignore the small amount of energy radiated as sound. Then the energy version of the isolated system model implies that the initial energy (kinetic) of the system of the car equals the final energy (elastic potential).

$$K_i = U_f \qquad \text{or} \qquad \tfrac{1}{2}mv^2 = \tfrac{1}{2}kx^2$$

$$v = \sqrt{\frac{k}{m}}x = \sqrt{\frac{5.00 \times 10^6 \text{ N / m}}{1000 \text{ kg}}}\,(3.16 \times 10^{-2} \text{ m}) = 2.23 \text{ m/s} \qquad \lozenge$$

---

21. A particle executes simple harmonic motion with an amplitude of 3.00 cm. At what position does its speed equal one half of its maximum speed?

Solution   If we consider the speed of the particle along its path as shown in the graphical and tabular representations, we can see that the particle is at rest momentarily at one endpoint while being accelerated toward the middle by an elastic force that decreases as the particle approaches the equilibrium position.

When it reaches the midpoint, the direction of acceleration changes so that the particle slows down until it stops momentarily at the opposite endpoint. From this analysis, we can estimate that $v = v_{max}/2$ somewhere in the outer half of the travel (since this is the region where the speed is changing most rapidly):

$$1.50 < \pm x < 3.00$$

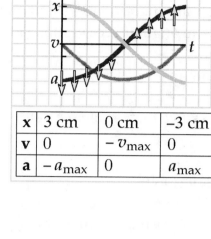

x	3 cm	0 cm	−3 cm
v	0	$-v_{max}$	0
a	$-a_{max}$	0	a_{max}

One way to analyze this problem is to start with the equation for linear SHM, take the first derivative with respect to time to find $v(t)$, and solve this equation for x when $v = v_{max}/2$.

For linear SHM,

$$x = A\cos\omega t$$

noting that the negative indicates direction,

$$dx/dt = v = -A\omega\sin\omega t$$

Since $A\omega$ is a constant, $v = \dfrac{v_{max}}{2}$ when

$$\sin\omega t = \pm\frac{1}{2}$$

or

$$\omega t = \sin^{-1}\left(\pm\frac{1}{2}\right) = \pm\frac{\pi}{6}$$

Thus, in our position equation,

$$\cos\omega t = \cos\left(\pm\frac{\pi}{6}\right) = \frac{\sqrt{3}}{2}$$

Substituting $A = 3.00$ cm,

$$x = \pm\frac{A\sqrt{3}}{2} = \pm\frac{(3.00\ \text{cm})\sqrt{3}}{2}$$

Solving,

$$x = \pm 2.60\ \text{cm} \qquad\qquad ◊$$

We could get the same answer from

$$v = \pm\omega\sqrt{A^2 - x^2}$$

The calculated position is in the outer half of the travel as predicted, and is in fact very close to the endpoints. This means that the speed of the particle changes remarkably little until the particle reaches the ends of its travel, where it experiences the maximum restoring force of the spring, which is proportional to x.

23. A simple pendulum has a mass of 0.250 kg and a length of 1.00 m. It is displaced through an angle of 15.0° and then released. What are (a) the maximum speed, (b) the maximum angular acceleration, and (c) the maximum restoring force?

Solution

We can solve this problem by either of two methods.

METHOD ONE:

Since 15.0° is small enough that (in radians) $\sin \theta \approx \theta$ within 1%, we may model the motion as simple harmonic motion. The constant angular frequency characterizing the motion is

$$\omega = \sqrt{\frac{g}{L}} = \sqrt{\frac{9.80 \text{ m/s}^2}{1.00 \text{ m}}} = 3.13 \text{ rad/s}$$

The amplitude as a distance is $\qquad A = L\theta = (1.00 \text{ m})(0.262) = 0.262 \text{ m}$

(a) The maximum linear speed is $\qquad v_{max} = \omega A = (3.13 \text{ s}^{-1})(0.262 \text{ m}) = 0.820 \text{ m / s}$ ◊

(b) Similarly, $\qquad a_{max} = \omega^2 A = (3.13 \text{ 1/s})^2 (0.262 \text{ m}) = 2.57 \text{ m/s}^2$

This implies maximum angular acceleration $\qquad \alpha = \dfrac{a}{r} = \dfrac{2.57 \text{ m/s}^2}{1.00 \text{ m}} = 2.57 \text{ rad/s}^2$ ◊

(c) $\Sigma F = ma = (0.250 \text{ kg})(0.257 \text{ m/s}^2) = 0.641 \text{ N}$ ◊

METHOD TWO:

We may work out slightly more precise answers by using the energy version of the isolated system model and the particle under a net force model. At release, the pendulum has height above its equilibrium position.

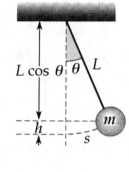

$$h = L - L \cos 15.0° = 1.00 \text{ m}(1 - \cos 15.0°) = 0.0341 \text{ m}$$

(a) The energy of the pendulum - Earth system is conserved as the pendulum swings down:

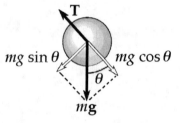

$$(K + U)_{top} = (K + U)_{bottom}$$

$$0 + mgh = \tfrac{1}{2} m v_{max}^2 + 0$$

$$v_{max} = \sqrt{2gh} = \sqrt{2(9.80 \text{ m}/\text{s}^2)0.0341 \text{ m}} = 0.817 \text{ m}/\text{s}$$

(c) The restoring force at release is

$$mg \sin 15.0° = (0.250 \text{ kg})(9.80 \text{ m}/\text{s}^2)\sin 15.0° = 0.634 \text{ N}$$

(b) This produces linear acceleration

$$a = \frac{\Sigma F}{m} = \frac{0.634 \text{N}}{0.250 \text{ kg}} = 2.54 \text{ m}/\text{s}^2$$

and angular acceleration

$$\alpha = \frac{a}{r} = \frac{2.54 \text{ m}/\text{s}^2}{1.00 \text{ m}} = 2.54 \text{ rad}/\text{s}^2$$

25. A particle of mass m slides inside a frictionless hemispherical bowl of radius R. Show that if it starts from rest with a small displacement from equilibrium, the particle moves in simple harmonic motion with an angular frequency equal to that of a simple pendulum of length R (i.e., $\omega = \sqrt{g/R}$).

Solution

Locate the center of curvature C of the bowl. We can measure the excursion of the object from equilibrium by the angle θ between the radial line to C and the vertical. The distance the object moves from equilibrium is $s = R\theta$.

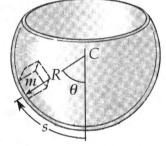

$\Sigma F_s = ma$ becomes

$$-mg \sin \theta = m \frac{d^2s}{dt^2}$$

For small angles

$$\theta \cong \sin \theta$$

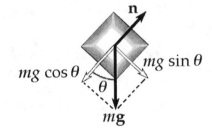

so by substitution,

$$-mg\theta = m \frac{d^2s}{dt^2}$$

$$-mg \frac{s}{R} = m \frac{d^2s}{dt^2}$$

Isolating the derivative,

$$\frac{d^2s}{dt^2} = -\left(\frac{g}{R}\right)s$$

By the form of this equation, we can see that the acceleration is proportional to the position and in the opposite direction, so we have SHM.   ◊

We identify its angular frequency by comparing our equation to Eq. 12.5:

$$\frac{d^2x}{dt^2} = -\omega^2 x$$

Now x and s both measure position, so $\omega^2 = \dfrac{g}{R}$ and $\omega = \sqrt{\dfrac{g}{R}}$ ◊

27. A physical pendulum in the form of a planar body moves in simple harmonic motion with a frequency of 0.450 Hz. If the pendulum has a mass of 2.20 kg and the pivot is located 0.350 m from the center of mass, determine the moment of inertia of the pendulum about the pivot point.

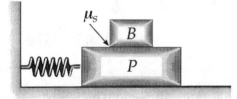

Solution

$$f = 0.450 \text{ Hz}, \qquad d = 0.350 \text{ m}, \quad \text{and} \qquad m = 2.20 \text{ kg}$$

Using equation 12.27, $\qquad T = 2\pi\sqrt{\dfrac{I}{mgd}} \qquad\qquad T^2 = \dfrac{4\pi^2 I}{mgd}$

$$I = \frac{T^2 mgd}{4\pi^2} = \left(\frac{1}{f}\right)^2 \frac{mgd}{4\pi^2} = \frac{(2.20 \text{ kg})(9.80 \text{ m / s}^2)(0.350 \text{ m})}{(0.450 \text{ s}^{-1})^2 (4\pi^2)} = 0.944 \text{ kg} \cdot \text{m}^2 \qquad \Diamond$$

---

39. A large block P executes horizontal simple harmonic motion as it slides across a frictionless surface with a frequency $f = 1.50$ Hz. Block B rests on it, as shown in Figure P12.39, and the coefficient of static friction between the two is $\mu_S = 0.600$. What maximum amplitude of oscillation can the system have if block B is not to slip?

Figure P12.39

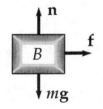

Solution　　If the block B does not slip, it undergoes simple harmonic motion (SHM) with the same amplitude and frequency as those of P, and with its acceleration caused by the static friction force exerted on it by P. Think of the block when it is just ready to slip at a turning point in its motion:

$\Sigma F = ma$ becomes $\qquad f_{\max} = \mu_s n = \mu_s mg = ma_{\max} = mA\omega^2$

Then $\qquad\qquad A = \dfrac{\mu_s g}{\omega^2} = \dfrac{0.600(9.80 \text{ m / s}^2)}{[2\pi(1.50 / \text{s})]^2} = 6.62 \text{ cm} \qquad \Diamond$

---

41. A pendulum of length L and mass M has a spring of force constant k connected to it at a distance h below its point of suspension (Fig. P12.41). Find the frequency of vibration of the system for small values of the amplitude (small θ). Assume the vertical suspension of length L is rigid but neglect its mass.

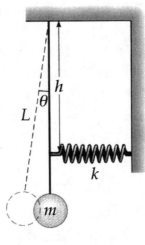

Solution   For the pendulum (see sketch), we use the rigid body under net torque model:

$$\Sigma \tau = I\alpha \quad \text{and} \quad \frac{d^2\theta}{dt^2} = -\alpha$$

The negative sign appears because positive θ is measured clockwise in the picture. We take torque around the point of suspension, so that the hinge force **H** contributes no torque.

Figure P12.41

$$\Sigma \tau = MgL\sin\theta + kxh\cos\theta = I\alpha$$

For small amplitude vibrations, use the approximations:

$$\sin\theta \approx \theta, \quad \cos\theta \approx 1, \quad \text{and} \quad x \approx s = h\theta$$

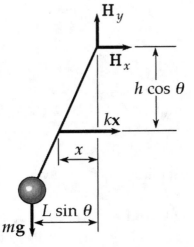

Therefore, with $I = mL^2$,

$$\frac{d^2\theta}{dt^2} = -\left(\frac{MgL + kh^2}{I}\right)\theta = -\left(\frac{MgL + kh^2}{ML^2}\right)\theta$$

This is of the form $\dfrac{d^2\theta}{dt^2} = -\omega^2\theta$ required for SHM,

with angular frequency, $\omega = \sqrt{\dfrac{MgL + kh^2}{ML^2}} = 2\pi f$

The frequency is $f = \dfrac{\omega}{2\pi} = \dfrac{1}{2\pi L}\sqrt{\dfrac{MgL + kh^2}{M}}$ ◊

49. A ball of mass m is connected to two rubber bands of length L, each under tension T, as in Figure P12.49. The ball is displaced by a small distance y perpendicular to the length of the rubber bands. Assuming the tension does not change, show that (a) the restoring force is $-(2T/L)y$ and (b) the system exhibits simple harmonic motion with an angular frequency $\omega = \sqrt{2T/mL}$.

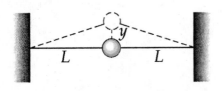

Figure P12.49

Solution

(a) $\Sigma \mathbf{F} = -2T \sin\theta \mathbf{j}$ where $\theta = \tan^{-1}\left(\dfrac{y}{L}\right)$

Since for a small displacement, $\sin\theta \approx \tan\theta = \dfrac{y}{L}$

and the resultant force is $\sum \mathbf{F} = \left(-\dfrac{2Ty}{L}\right)\mathbf{j}$ ◊

(b) Since there is a restoring force that is proportional to the position, it causes the system to move with simple harmonic motion like a block-spring system. Thus,

$\sum F = -kx$ becomes $\sum F = -\left(\dfrac{2T}{L}\right)y$

Therefore, $\omega = \sqrt{\dfrac{k}{m}} = \sqrt{\dfrac{2T}{mL}}$ ◊

Chapter 13

Mechanical Waves

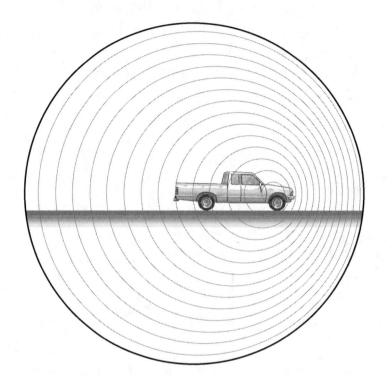

INTRODUCTION

In this chapter we study the properties of mechanical waves. In the case of mechanical waves, what we interpret as a wave corresponds to the disturbance of a body or medium through which the wave travels. Therefore, we can consider a wave to be the **motion of a disturbance.**

The mathematics used to describe wave phenomena is common to all waves. In general, we shall find that mechanical wave motion is described by specifying the positions of all points of the disturbed medium as a function of time.

Sound waves are important examples of longitudinal waves. They can travel through any material medium with a speed that depends on the properties of the medium. As the waves travel, the particles in the medium vibrate to produce density and pressure changes along the direction of motion of the wave. These changes result in a series of high- and low-pressure regions called **compressions** and **rarefactions,** respectively. We shall find that the mathematical description of harmonic sound waves is identical to that of harmonic string waves.

NOTES FROM SELECTED CHAPTER SECTIONS

13.1 Propagation of a Disturbance

A mechanical wave requires:

(1) a source of disturbance

(2) a medium which can be disturbed

(3) a physical mechanism through which particles of the medium can influence one another.

The wave function $y(x,t)$ represents a one-dimensional wave and gives the y coordinate of any point P located at position x at any time t.

A transverse wave is one in which the particles of the disturbed medium oscillate back and forth along a direction perpendicular to the direction of the wave velocity.

A longitudinal wave is one in which the particles of the medium undergo a displacement (oscillate back and forth) along a direction parallel to the direction of the wave velocity (direction along which the wave travels).

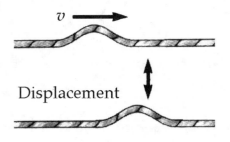

Transverse wave   Longitudinal wave

A one-dimensional traveling wave can be described mathematically by its **wave function** $y(x,t) = f(x \mp vt)$. If the wave is assumed to be traveling along the x direction, then $y(x, t)$ represents the y coordinate of any point on the string at any time, t.

In the expression for the wave function, the **negative sign** describes a wave traveling **toward the right** and the **positive sign** describes the wave function for a wave traveling **toward the left**.

In the case of a wave on a string at a **fixed value of** x, the wave function represents the y coordinate of a particular **point as a function of time**. On the other hand, if t **is fixed**, the wave function defines a curve showing the **shape of the wave pulse** at a given time.

13.2   The Wave Model

There are three parameters important in characterizing waves:

- **Wavelength** is the minimum distance between any two points on a wave that behave identically.

- **Frequency** is the rate at which the disturbance of the elastic medium repeats.

- **Speed** is the rate at which the wave propagates through the medium.

A sinusoidal wave is one whose shape is sinusoidal at every instant of time. Sinusoidal waves which differ in phase are shown in the adjacent figure.

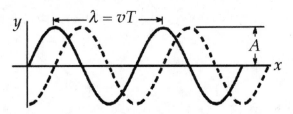

The wave amplitude, A, is the maximum possible value of the position of a particle of the medium relative to its equilibrium position.

The wavelength, λ, is the minimum distance between two points which are the same distance from their equilibrium positions and are moving in the same direction (such a pair of points are said to be in phase).

The period of the wave is the time required for the disturbance to travel along the direction of propagation a distance equal to the wavelength. The period is also the time required for any point in the medium to complete one complete cycle in its harmonic motion about its equilibrium point.

13.4 The Speed of Transverse Waves on Strings

For linear waves, the **speed** of **mechanical waves** depends only on the physical properties of the medium through which the disturbance travels. In the case of waves on a **string**, the velocity depends on the tension in the string and the mass per unit length (linear mass density).

13.5 Reflection and Transmission of Waves

The following general rules apply to reflected waves: **When a wave pulse travels from medium A to medium B and $v_A > v_B$ (that is, when B is more dense than A), the reflected part of the pulse is inverted upon reflection. When a wave pulse travels from medium A to medium B and $v_A < v_B$ (A is more dense than B), the reflected pulse is not inverted.**

13.6 Rate of Energy Transfer by Sinusoidal Waves on Strings

As waves propagate through a medium, they transport energy; this occurs without any net transfer of matter. The power transmitted by any sinusoidal wave is proportional to the square of the angular frequency and to the square of the amplitude.

13.7 Sound Waves

The motion of the parcels of the medium is **back and forth along the direction in which the sound wave travels.** This is in contrast to a transverse wave, in which the vibrations of the medium are **at right angles to the direction of travel of the wave.** The **pressure amplitude** is proportional to the **displacement amplitude**; however, the pressure wave is 90° out of phase with the displacement.

13.8 The Doppler Effect

In general, a Doppler effect is experienced whenever there is relative motion between source and observer. When the source and observer are moving toward each other, the frequency heard by the observer is higher than the frequency of the source. When the source and observer are moving away from each other, the observer hears a frequency lower than the source frequency.

EQUATIONS AND CONCEPTS

The value of the position repeats when x is increased by an integral multiple of λ and the wave moves to the right a distance of vt in a time t.

$$y = A\sin\left(\frac{2\pi}{\lambda}(x - vt)\right) \qquad (13.4)$$

The period T is the time it takes the wave to travel a distance of one wavelength.

$$v = \frac{\lambda}{T} \qquad (13.5)$$

It is convenient to define three additional characteristic wave quantities: **the wave number** k, the **angular frequency** ω, and the **frequency** f.

$$k \equiv \frac{2\pi}{\lambda} \qquad (13.7)$$

$$\omega \equiv \frac{2\pi}{T} = 2\pi f \qquad (13.8)$$

$$f = \frac{1}{T}$$

The expression for the **wave function** can be written in a more compact form in terms of the parameters defined above. If the transverse position is not zero at $x = 0$ and $t = 0$, it is necessary to include a phase constant, ϕ.

$$y = A\sin(kx - \omega t) \qquad (13.9)$$

$$kx - \omega t + \phi$$

$$y = A\sin(kx - \omega t + \phi) \qquad (13.12)$$

$$\cos(\omega t + \phi)$$

$$\sin(kx - \omega t + \phi)$$

The **wave speed** v or phase velocity can also be expressed in alternative forms.

$$v = \frac{\omega}{k} \qquad (13.10)$$

$$v = \lambda f \qquad (13.11)$$

$$A\cos(\omega t + \phi)$$

The **transverse velocity** v_y of a point on a harmonic wave is out of phase with the **transverse acceleration** a_y of that point by $\pi/2$ radians.

$$v_y = -\omega A \cos(kx - \omega t) \qquad (13.13)$$

$$a_y = -\omega^2 A \sin(kx - \omega t) \qquad (13.14)$$

In the special case of a transverse pulse moving along a stretched string, the wave speed depends on the tension in the string T and the linear density μ of the string (mass per unit length).

$$v = \sqrt{\frac{T}{\mu}} \qquad (13.20)$$

The **power** transmitted by any harmonic wave is proportional to the square of the frequency and the square of the amplitude, where μ is the mass per unit length of the string.

$$\mathcal{P} = \frac{1}{2}\mu\omega^2 A^2 v \qquad (13.23)$$

A harmonic sound wave is produced in a gas when the source is a body vibrating in simple harmonic motion (such as a vibrating guitar string). Under these conditions, both the **displacement** of the medium s and the **pressure variations** of the gas ΔP vary harmonically in time. Note that the displacement and the pressure variation are out of phase by $\pi/2$.

$$s(x,\ t) = s_{\max} \sin(kx - \omega t) \qquad (13.24)$$

$$\Delta P = \Delta P_{\max} \cos(kx - \omega t) \qquad (13.25)$$

A sound wave may be considered as either a displacement wave or a pressure wave. The **pressure amplitude** is **proportional** to the **displacement amplitude.**

$$\Delta P_{\max} = \rho v \omega s_{\max} \qquad (13.26)$$

The change in frequency heard by an observer whenever there is **relative motion between the source and the observer** is called the **Doppler effect**. In Equation 13.30, v_o and v_s are measured **relative to the medium in which the sound travels.**

$$f' = f\left(\frac{v + v_o}{v - v_s}\right) \qquad (13.30)$$

In the general equation for the observed frequency (Eq. 13.30), a **positive** sign is used for motion of the observer (v_o) or source (v_s) toward the other and a **negative** sign is used for motion of one away from the other. You should confirm that this rule gives the results shown in the following table.

Observer	Source	Equation	Remark
O →	S	$f' = f\left(\dfrac{v + v_o}{v}\right)$	Observer moving toward stationary source
← O	S	$f' = f\left(\dfrac{v - v_o}{v}\right)$	Observer moving away from stationary source
O	← S	$f' = f\left(\dfrac{v}{v - v_s}\right)$	Source moving toward stationary observer
O	S →	$f' = f\left(\dfrac{v}{v + v_s}\right)$	Source moving away from stationary observer
O →	S →	$f' = f\left(\dfrac{v + v_o}{v + v_s}\right)$	Observer following moving source
← O	← S	$f' = f\left(\dfrac{v - v_o}{v - v_s}\right)$	Source following moving observer
← O	S →	$f' = f\left(\dfrac{v - v_o}{v + v_s}\right)$	Observer and source both moving away from each other
O →	← S	$f' = f\left(\dfrac{v + v_o}{v - v_s}\right)$	Observer and source moving toward each other
O	S	$f' = f$	Observer and source both stationary

EXAMPLES OF DOPPLER EFFECT WITH OBSERVER/SOURCE IN MOTION

REVIEW CHECKLIST

▷ Recognize whether or not a given function is a possible description of a traveling wave.

▷ Express a given harmonic wave function in several alternative forms involving different combinations of the wave parameters: wavelength, period, phase velocity, wave number, angular frequency, and harmonic frequency.

▷ Given a specific wave function for a harmonic wave, obtain values for the characteristic wave parameters: A, ω, k, λ, f, and ϕ.

▷ Calculate the rate at which energy is transported by harmonic waves in a string.

▷ Describe the harmonic displacement and pressure variation as functions of time and position for a harmonic sound wave. Relate the displacement amplitude to the pressure amplitude for a harmonic sound wave.

▷ Describe the various situations under which a Doppler shifted frequency is produced. Note that a Doppler shift is observed as long as there is a **relative** motion between the observer and the source.

ANSWERS TO SELECTED CONCEPTUAL QUESTIONS

2. How would you set up a longitudinal wave in a stretched spring? Would it be possible to set up a transverse wave in a spring?

Answer   A longitudinal wave can be set up in a stretched spring by compressing the coils in a small region, and releasing the compressed region. The disturbance will proceed to propagate as a longitudinal pulse. It is quite possible to set up a transverse wave in a spring, simply by displacing the spring in a direction perpendicular to its length.

Chapter 13

7. A vibrating source generates a sinusoidal wave on a string under constant tension. If the power delivered to the string is doubled, by what factor does the amplitude change? Does the wave speed change under these circumstances?

Answer Power is always proportional to the square of the amplitude, so if power doubles, amplitude increases by $\sqrt{2}$ times, a factor of 1.41. The wave speed does not depend on amplitude, but stays constant.

□ □ □ □

13. If you stretch a rubber hose and pluck it, you can observe a pulse traveling up and down the hose. What happens to the speed if you stretch the hose tighter? If you fill the hose with water?

Answer If you stretch the hose tighter, you increase the tension, and increase the speed of the wave. If you fill it with water, you increase the linear density of the hose, and decrease the speed of the wave.

□ □ □ □

SOLUTIONS TO SELECTED END-OF-CHAPTER PROBLEMS

1. At $t = 0$, a transverse wave pulse in a wire is described by the function

$$y = \frac{6}{x^2 + 3}$$

where x and y are in meters. Write the function $y(x, t)$ that describes this wave if it is traveling in the positive x direction with a speed of 4.50 m/s.

236

Solution $y(x, t)$ must be a function of both x and t, but must become $y = 6/(x^2 + 3)$ when $t = 0$. To guarantee the same form, we substitute the term $x = x' + ut$, and solve for y:

$$y = \frac{6}{(x' + ut)^2 + 3}$$

Note that as t increases, x' must decrease by $u\Delta t$; aside from that, the equation remains the same. We first define that at $t = 0$, $x = 0$.

So $x' + u(0) = x = 0$ (1)

In order to cause the wave to appear to move to the right, we need to force our reference point (x) to the left. Therefore, 1 second later, at $t = 1$, the wave has moved 4.50 m in the $+x$ direction, but x moves 4.50 m in the $-x$ direction.

$$x = x' + u(1) = -4.50 \text{ m}$$ (2)

Subtracting equations (1) and (2), $u = -4.50$ m/s, and our new equation is:

$$y(x, t) = \frac{6}{(x - 4.50t)^2 + 3}$$ ◊

In general, we can cause any waveform to move along the x axis at a velocity v_x by substituting $(x - v_x t)$ for x. The same principle applies to motion in other directions.

---

3.   A sinusoidal wave is traveling along a rope. The oscillator that generates the wave completes 40.0 vibrations in 30.0 s. Also, a given maximum travels 425 cm along the rope in 10.0 s. What is the wavelength?

Solution $f = \dfrac{40.0 \text{ waves}}{30.0 \text{ s}} = 1.33 \text{ s}^{-1}$ and $v = \dfrac{425 \text{ cm}}{10.0 \text{ s}} = 42.5 \text{ cm / s}$

Since $v = \lambda f$, $\lambda = \dfrac{v}{f} = \dfrac{42.5 \text{ cm / s}}{1.33 \text{ s}^{-1}} = 0.319 \text{ m}$ ◊

---

5.   The wave function for a traveling wave on a taut string is (in SI units)

$$y(x,\ t) = (0.350\ \text{m})\ \sin\ (10\pi t - 3\pi x + \pi/4)$$

(a) What are the speed and direction of travel of the wave?   (b) What is the vertical displacement of the string at $t = 0$, $x = 0.100$ m?   (c) What are the wavelength and frequency of the wave?   (d) What is the maximum magnitude of the transverse speed of the string?

Solution   We use the traveling wave model.

We compare the given equation with $y = A \sin (kx - \omega t + \phi)$; and find that

$$k = 3\pi\ \text{rad/m} \qquad\qquad \text{and} \qquad\qquad \omega = 10\pi\ \text{rad/s}$$

(a)   The speed and direction of the wave can be defined in terms of the velocity:

$$\mathbf{v} = f\lambda\,\mathbf{i} = \frac{\omega}{k}\,\mathbf{i} = \frac{10\pi\ \text{rad/s}}{3\pi\ \text{rad/m}}\,\mathbf{i} = 3.33\mathbf{i}\ \text{m/s} \qquad\qquad \Diamond$$

(b)   Substituting $t = 0$ and $x = 0.100$ m,

$$y = (0.350\ \text{m})\ \sin\ (-0.300\pi + 0.250\pi) = -0.0548\ \text{m} = -5.48\ \text{cm} \qquad\qquad \Diamond$$

Note that when you take the sine of a quantity with no units, it is not in degrees, but in radians.

(c) $$\lambda = \frac{2\pi\ \text{rad}}{k} = \frac{2\pi\ \text{rad}}{3\pi\ \text{rad/m}} = 0.667\ \text{m} \qquad\qquad \Diamond$$

and $$f = \frac{\omega}{2\pi\ \text{rad}} = \frac{10\pi\ \text{rad/s}}{2\pi\ \text{rad}} = 5.00\ \text{Hz} \qquad\qquad \Diamond$$

(d) $$v_y = \frac{\partial y}{\partial t} = (0.350\ \text{m})(10\pi\ \text{rad/s})\ \cos\ (10\pi t - 3\pi x + \pi/4) \qquad\qquad \Diamond$$

The maximum occurs when the cosine term is 1:

$$v_{y,\text{max}} = (10\pi\ \text{rad/s})(0.350\ \text{m/s}) = 11.0\ \text{m/s} \qquad\qquad \Diamond$$

Note the large difference between the maximum particle speed and the wave speed found in part (a).

9. (a) Write the expression for y as a function of x and t for a sinusoidal wave traveling along a rope in the **negative** x direction with the following characteristics: $A = 8.00$ cm, $\lambda = 80.0$ cm, $f = 3.00$ Hz, and $y(0, t) = 0$ at $t = 0$. (b) Write the expression for y as a function of x and t for the wave in (a) assuming that $y(x, 0) = 0$ at the point $x = 10.0$ cm.

Solution The amplitude is $\qquad A = y_{max} = 8.00$ cm $= 0.0800$ m

$$k = \frac{2\pi}{\lambda} = \frac{2\pi}{0.800 \text{ m}} = 7.85 \text{ m}^{-1} \qquad \omega = 2\pi f = 2\pi(3.00 \text{ s}^{-1}) = 6.00\pi \text{ rad / s}$$

(a) Since $\phi = 0$, $\qquad\qquad\qquad\qquad y = A \sin (kx + \omega t + \phi)$

becomes $\qquad\qquad\qquad\qquad\qquad y = (0.0800 \text{ m}) \sin (7.85x + 6.00\pi t)$ ◊

(b) In general, $\qquad\qquad\qquad\qquad y = (0.0800 \text{ m}) \sin (7.85x + 6.00\pi t + \phi)$.

If $y (x, 0) = 0$ at $x = 0.100$ m, $\qquad 0 = (0.0800 \text{ m}) \sin (0.785 + \phi)$ and $\phi = -0.785$ rad

Therefore, $\qquad\qquad\qquad\qquad y = (0.0800 \text{ m}) \sin (7.85x + 6.00\pi t - 0.785)$ ◊

---

15. A 30.0-m steel wire and a 20.0-m copper wire, both with 1.00-mm diameters, are connected end to end and stretched to a tension of 150 N. How long does it take a transverse wave to travel the entire length of the two wires?

Solution The total time of travel is the sum of the two times.

In each wire, $t = \dfrac{L}{v} = L\sqrt{\dfrac{\mu}{T}}$ where $\mu = \dfrac{m}{L} = \rho\dfrac{V}{L} = \rho A = \dfrac{\pi\rho d^2}{4}$, so $t = L\sqrt{\dfrac{\pi\rho d^2}{4T}}$

For copper, $t_1 = (20.0 \text{ m})\sqrt{\dfrac{\pi(8920 \text{ kg / m}^3)(0.00100 \text{ m})^2}{4(150 \text{ kg} \cdot \text{m / s}^2)}} = 0.137$ s

For steel, $t_2 = (30.0 \text{ m})\sqrt{\dfrac{\pi(7860 \text{ kg / m}^3)(0.00100 \text{ m})^2}{4(150 \text{ kg} \cdot \text{m / s}^2)}} = 0.192$ s

The total time is $\qquad (0.137 \text{ s}) + (0.192 \text{ s}) = 0.329$ s ◊

---

19. Sinusoidal waves 5.00 cm in amplitude are to be transmitted along a string that has a linear mass density of 4.00×10^{-2} kg/m. If the source can deliver a maximum power of 300 W and the string is under a tension of 100 N, what is the highest vibrational frequency at which the source can operate?

Solution   The wave speed $v = \sqrt{\dfrac{T}{\mu}} = \sqrt{\dfrac{100 \text{ N}}{4.00 \times 10^{-2} \text{ kg/m}}} = 50.0$ m/s

From $\mathscr{P} = \frac{1}{2}\mu\omega^2 A^2 v,$ $\omega^2 = \dfrac{2\mathscr{P}}{\mu A^2 v} = \dfrac{2(300 \text{ N} \cdot \text{m/s})}{\left(4.00 \times 10^{-2} \text{ kg/m}\right)\left(5.00 \times 10^{-2} \text{ m}\right)^2 (50.0 \text{ m/s})}$

Solving, $\omega = 346.4$ rad/s   and $f = \dfrac{\omega}{2\pi} = 55.1$ Hz   ◊

---

23. Suppose that you hear a clap of thunder 16.2 s after seeing the associated lightning stroke. The speed of sound waves in air is 343 m/s, and the speed of light in air is 3.00×10^8 m/s. How far are you from the lightning stroke?

Solution

There is a common rule of thumb that lightning is about a mile away for every 5 seconds of delay between the flash and thunder (or ~ 3 s/km). Therefore, this lightning strike is about 3 miles (~ 5 km) away.

The distance can be found from the speed of sound and the elapsed time. The time for the light to travel to the observer will be much less than the sound delay, so the speed of light can be taken as ∞.

Assuming that the speed of sound is constant through the air between the lightning strike and the observer,

$v_s = \dfrac{d}{\Delta t}$ or $d = v_s \Delta t = (343 \text{ m/s})(16.2 \text{ s}) = 5.56$ km   ◊

Our calculated answer is consistent with our initial estimate, but we should check the validity of our assumption that the speed of light could be ignored. The time delay for the light is

$$t_{light} = \frac{d}{c_{air}} = \frac{5560 \text{ m}}{3.00 \times 10^8 \text{ m / s}} = 1.85 \times 10^{-5} \text{ s}$$

and $\qquad \Delta t = t_{sound} - t_{light} = 16.2 \text{ s} - 1.85 \times 10^{-5} \text{ s} \approx 16.2 \text{ s (when rounded)}$

Since the travel time for the light is much smaller than the uncertainty in the time of 16.2 s, t_{light} can be ignored without affecting the distance calculation.

However, our assumption of a constant speed of sound in air is probably not valid due to local variations in air temperature during a storm. We must assume that the given speed of sound in air is an accurate **average** value for the conditions described.

---

27. An experimenter wishes to generate in air a sound wave that has a displacement amplitude of 5.50×10^{-6} m. The pressure amplitude is to be limited to 0.840 N/m^2. What is the minimum wavelength the sound wave can have?

Solution

We are given $s_{max} = 5.50 \times 10^{-6}$ m and $\Delta P_{max} = 0.840$ Pa. The pressure amplitude is

$$\Delta P_{max} = \rho v \omega s_{max} = \rho v \left(\frac{2\pi v}{\lambda} \right) s_{max}$$

or $\qquad \lambda_{min} = \frac{2\pi \rho v^2 s_{max}}{\Delta P_{max}} = \frac{2\pi (1.20 \text{ kg/m}^3)(343 \text{ m/s})^2 (5.50 \times 10^{-6} \text{ m})}{0.840 \text{ Pa}} = 5.81 \text{ m} \qquad \lozenge$

---

29. Write an expression that describes the pressure variation as a function of position and time for a sinusoidal sound wave in air if $\lambda = 0.100$ m and $\Delta P_{max} = 0.200$ Pa.

Solution We write the pressure variation as $\Delta P = \Delta P_{max} \sin(kx - \omega t)$.

Noting that $k = \dfrac{2\pi}{\lambda}$, $\qquad\qquad k = \dfrac{2\pi \text{ rad}}{0.100 \text{ m}} = 62.8 \text{ rad / m}$

Likewise, $\omega = \dfrac{2\pi v}{\lambda}$, so $\qquad \omega = \dfrac{(2\pi \text{ rad})(343 \text{ m / s})}{0.100 \text{ m}} = 2.16 \times 10^4 \text{ rad / s}$

We now can create our equation: $\qquad \Delta P = (0.200 \text{ Pa}) \sin(62.8x - 21600t)$ ◊

---

31. Standing at a crosswalk, you hear a frequency of 560 Hz from the siren of an approaching ambulance. After the ambulance passes, the observed frequency of the siren is 480 Hz. Determine the ambulance's speed from these observations.

Solution

For an approaching car (Eq. 13.29): $\qquad\qquad f' = \dfrac{f}{\left(1 - \dfrac{v_s}{v}\right)}$

For a departing car: $\qquad\qquad f'' = \dfrac{f}{\left(1 + \dfrac{v_s}{v}\right)}$

Since $f' = 560$ Hz and $f'' = 480$ Hz, $\qquad 560\left(1 - \dfrac{v_s}{v}\right) = 480\left(1 + \dfrac{v_s}{v}\right)$

$1040 \dfrac{v_s}{v} = 80$ $\qquad\qquad v_s = \dfrac{80(343)}{1040} \text{ m / s} = 26.4 \text{ m / s}$ ◊

33. A tuning fork vibrating at 512 Hz falls from rest and accelerates at 9.80 m/s². How far below the point of release is the tuning fork when waves of frequency of 485 Hz reach the release point? Take the speed of sound in air to be 340 m/s.

Solution   In order to solve this problem, we must first determine how fast the tuning fork is falling when its frequency is 485 Hz.

The tuning fork (source) is moving **away** from a stationary listener.

Therefore, we use the equation $\qquad f' = \left(\dfrac{v}{v + v_s}\right)f:$

$$485\ \text{Hz} = (512\ \text{Hz})\frac{340\ \text{m}/\text{s}}{340\ \text{m}/\text{s} + v_{\text{fall}}}$$

Solving, $\qquad v_{\text{fall}} = (340\ \text{m}/\text{s})\left(\frac{512\ \text{Hz}}{485\ \text{Hz}} - 1\right) = 18.93\ \text{m}/\text{s}$

For the tuning fork we use the particle under constant acceleration model. From the kinematic equation $v_2{}^2 = v_1{}^2 + 2as$, we calculate that

$$s = \frac{v_2{}^2}{2a} = \frac{(18.93\ \text{m}/\text{s})^2}{2(9.80\ \text{m}/\text{s}^2)} = 18.28\ \text{m}$$

Since $v = at$, $\qquad t = \dfrac{v_{\text{fall}}}{a} = \dfrac{18.93\ \text{m}/\text{s}}{9.80\ \text{m}/\text{s}^2} = 1.931\ \text{s}$

At this moment, the fork would appear to ring at 485 Hz **just below the fork**. However, it takes some additional time for the waves to reach the point of release. From the traveling wave model,

$$\Delta t = \frac{s}{v} = \frac{18.28\ \text{m}}{340\ \text{m}/\text{s}} = 0.0538\ \text{s}$$

Over the total time $t + \Delta t$, the fork falls a distance

$$s_{\text{total}} = \tfrac{1}{2}a(t + \Delta t)^2 = \tfrac{1}{2}(9.80\ \text{m}/\text{s}^2)(1.985\ \text{s})^2 = 19.3\ \text{m} \qquad \Diamond$$

45. A rope of total mass m and length L is suspended vertically. Show that a transverse wave pulse will travel the length of the rope in a time $t = 2\sqrt{L/g}$. (**Hint:** First find an expression for the wave speed at any point a distance x from the lower end by considering the tension in the rope as resulting from the weight of the segment below that point.)

Solution

We define $x = 0$ at the bottom of the rope and $x = L$ at the top of the rope. The tension in the rope at any point is the weight of the rope below that point. We can thus write the tension in the rope at each point x as $F = \mu x g$, where μ is the mass per unit length of the rope.

The speed of the wave pulse at each point along the rope's length is therefore

$$v = \sqrt{\frac{F}{\mu}} \qquad \text{or} \qquad v = \sqrt{gx}$$

But at each point x, the wave phase progresses at a rate of $v = \dfrac{dx}{dt}$.

So we can then substitute for v, and generate the differential equation:

$$\frac{dx}{dt} = \sqrt{gx} \qquad\qquad dt = \frac{dx}{\sqrt{gx}}$$

Integrating both sides, $\qquad t = \dfrac{1}{\sqrt{g}}\displaystyle\int_0^L \frac{dx}{\sqrt{x}} = \dfrac{\left[2\sqrt{x}\right]_0^L}{\sqrt{g}} = 2\sqrt{\dfrac{L}{g}}$ $\qquad\qquad\qquad \Diamond$

Chapter 14

Superposition and Standing Waves

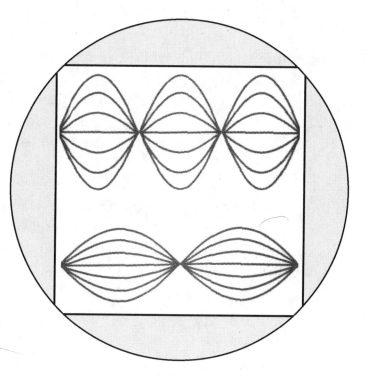

INTRODUCTION

An important aspect of waves is the combined effect of two or more of them traveling in the same medium.

In a linear medium, that is, one in which the restoring force on a particle of the medium is proportional to the displacement of the particle from equilibrium, the principle of superposition can be applied to obtain the resultant disturbance.

This chapter is concerned with the superposition principle as it applies to sinusoidal waves. If the sinusoidal waves that combine in a given medium have the same frequency and wavelength, and are traveling in opposite directions, one finds that a stationary pattern, called a **standing wave,** can be produced at certain frequencies.

NOTES FROM SELECTED CHAPTER SECTIONS

14.1   The Principle of Superposition

The **superposition principle** states that, when two or more waves move in the same linear medium, the **net displacement** of the medium at any point (the resultant wave) equals the algebraic sum of the displacements of all the waves. If the individual waves are sinusoidal and of equal frequency, the resultant wave function is also sinusoidal and has the **same frequency** and **same wavelength** as the individual waves.

Figure 14.1 shows the resultant of two traveling sinusoidal waves for
(a) $\phi = 0, 2\pi, 4\pi, \ldots$ corresponding to constructive interference,
(b) $\phi = \pi, 3\pi, 5\pi, \ldots$ corresponding to destructive interference, and
(c) $0 < \phi < \pi$ for which the resultant amplitude has a value between 0 and $2A_0$.

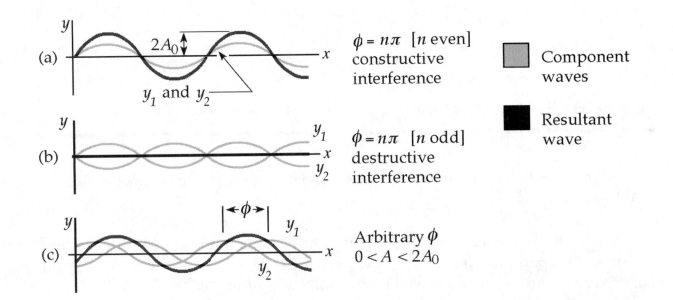

Figure 14.1   Superposition of two waves with the same amplitude and frequency, with a phase difference ϕ of (a) $n\pi$ (n even), (b) $n\pi$ (n odd), and (c) arbitrary ϕ. The gray curves represent y_1 and y_2; the dark curves represent $y = y_1 + y_2$.

14.4 Standing Waves in Strings

Standing waves can be set up in a string by a continuous superposition of waves incident on and reflected from the ends of the string. The string has a number of natural patterns or frequencies of vibration, called **normal modes** or **harmonics**. Each harmonic has a **characteristic frequency**. The lowest of these frequencies is called the **fundamental frequency**, which together with the higher frequencies form a **harmonic series**.

Figure 14.2 is a schematic representation of the first three normal modes of vibration of string fixed at both ends.

14.5 Standing Waves in Air Columns

Standing waves are produced in strings by interfering **transverse** waves. Sound sources can be used to produce **longitudinal** standing waves in air columns. The phase relationship between incident and reflected waves depends on whether or not the reflecting end of the air column is open or closed. This gives rise to two sets of possible standing wave conditions.

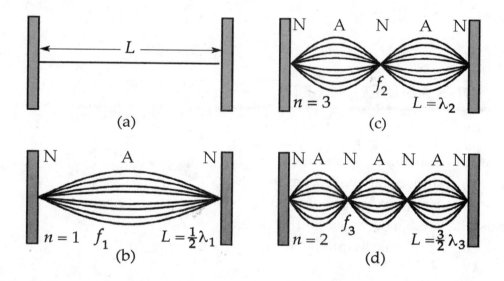

Figure 14.2 Schematic representation of standing waves on a stretched string of length L, where the envelope represents many successive vibrations. The points of zero displacement are called **nodes**; the points of maximum displacement are called **antinodes**. The pattern shown in (b) represents the first harmonic or fundamental frequency.

The first three natural modes of vibration for (a) an open pipe and (b) a closed pipe are shown in Figure 14.3.

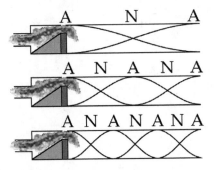 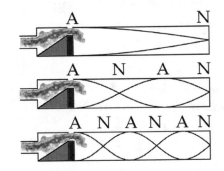

Figure 14.3 (a) Natural modes of vibration in a hollow pipe open at each end. All harmonics are present. **(b)** Natural modes of vibration for air in a hollow pipe closed at one end. Only odd harmonics are present.

EQUATIONS AND CONCEPTS

The wave function which is the resultant of two traveling sinusoidal waves having the same direction, frequency, and amplitude is also sinusoidal and has the same frequency and wavelength as the individual waves.

$$y = \left[2A\cos\frac{\phi}{2}\right]\sin\left(kx - \omega t + \frac{\phi}{2}\right) \qquad (14.1)$$

$$y \left(2A\cos\frac{\phi}{2}\right)\sin\left(kx - \omega t + \frac{\phi}{2}\right)$$

$$\|$$

$$\vee$$

The amplitude of the resultant wave depends on the phase difference between the two individual waves.

$$y_{max} = 2A\cos\frac{\phi}{2}$$

A **phase difference** can arise between two waves generated by the same source and arriving at a common point after having traveled along paths of **unequal path length.** In Equation 14.2, ϕ is the phase difference and Δr is the **path difference** between the two waves.

$$\Delta r = \frac{\lambda}{2\pi}\phi \qquad \text{(14.2)}$$

phase diff

path dif.

A **standing wave** can be produced in a string due to the interference of two sinusoidal waves with equal amplitude and frequency traveling in opposite directions.

$$y = (2A\sin kx)\cos\omega t \qquad \text{(14.3)}$$

$2A \sin kx \times \text{constant}$

The amplitude of a standing wave is a function of the position x along the string. **Antinodes** (A) are points of maximum displacement, while **nodes** (N) are points of zero displacement.

distance $\text{N to N} = \lambda/2$

$\text{A to A} = \lambda/2$

$\text{N to A} = \lambda/4$

A series of natural patterns of vibration called **normal modes** can be excited in a string fixed at both ends. Each mode corresponds to a characteristic frequency and wavelength. In Equation 14.8, T is the tension in the string and μ is the linear density.

$$\lambda_n = \frac{2L}{n} \qquad \text{(14.6)}$$

$$f_n = \frac{n}{2L}v \qquad \text{(14.7)}$$

$$f_n = \frac{n}{2L}\sqrt{\frac{T}{\mu}} \qquad n = 1, 2, 3, \ldots \qquad \text{(14.8)}$$

In an "**open**" pipe (open at both ends), all integral multiples of the fundamental frequency can be excited.

$$f_n = n\frac{v}{2L} \qquad n = 1, 2, 3, \ldots \qquad \text{(14.11)}$$

antinode $x = \frac{\lambda}{4} n \quad n = 1, 3, 5$

node $x = \frac{\lambda}{2} n \quad n = 1, 2, 3$

In a "**closed**" pipe (closed at one end), only the odd multiples of the fundamental frequency — the odd harmonics — are possible.

$$f_n = n\frac{v}{4L} \qquad n = 1, 3, 5, \ldots \qquad (14.12)$$

Beats are formed by the combination of two waves of equal amplitude but slightly different frequencies combining at one point in space. The resultant wave is simple harmonic with an effective frequency equal to the average of the two individual wave frequencies.

$$y = \left[2A\cos 2\pi\left(\frac{f_1 - f_2}{2}\right)t\right]\cos 2\pi\left(\frac{f_1 + f_2}{2}\right)t$$

$$(14.13)$$

The amplitude of the wave described by Equation 14.13 is time dependent. Each occurrence of maximum amplitude results in a "beat"; the **beat frequency** f_b equals the difference in the frequencies of the individual waves.

$$A_{x=0} = 2A\cos\left(2\pi\left(\frac{f_1 - f_2}{2}\right)t\right) \qquad (14.14)$$

$$f_b = |f_1 - f_2| \qquad (14.15)$$

Any **complex periodic wave form**, with a period T, can be represented by the combination of sinusoidal waves which form a harmonic series (combination of fundamental and various harmonics). Such a sum of sine and cosine terms is called a **Fourier series**. The lowest frequency is $f_1 = 1/T$ and higher frequencies are $f_n = nf_1$. A_n and B_n represent the amplitudes of the various harmonics.

$$y(t) = \sum_n \left(A_n \sin 2\pi f_n t + B_n \cos 2\pi f_n t\right) \quad (14.16)$$

REVIEW CHECKLIST

▷ Write out the wave function which represents the superposition of the two sinusoidal waves of equal amplitude and frequency traveling in opposite directions in the same medium.

▷ Given an expression for the wave function, identify the angular frequency and maximum amplitude. Determine the values of x which correspond to nodal and antinodal points of a standing wave.

▷ Calculate the normal mode frequencies for a string under tension, and for open and closed air columns.

▷ Describe the time dependent amplitude and determine the effective frequency of vibration when two waves of slightly different frequency interfere. Also, calculate the expected beat frequency for this situation.

ANSWERS TO SELECTED CONCEPTUAL QUESTIONS

1. When two waves interfere constructively or destructively, is there any gain or loss in energy? Explain.

Answer

No. The energy may be transformed into other forms of energy. For example, when two pulses traveling on a stretched string in opposite directions overlap, and one is inverted, some potential energy is transferred to kinetic energy when they overlap. In fact, if they have equal amplitudes in opposite directions, they completely cancel each other at one point. In this case, all of the energy is transverse kinetic energy when the resultant amplitude is zero.

2. Does the phenomenon of wave interference apply only to sinusoidal waves?

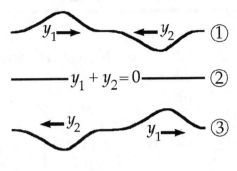

Answer

No. Any waves moving in the same medium can interfere with each other. For example, two pulses moving in opposite directions on a stretched string interfere when they meet each other.

□ □ □ □

10. An airplane mechanic notices that the sound from a twin-engine aircraft rapidly varies in loudness when both engines are running. What could be causing this variation from loud to soft?

Answer

Apparently the two engines are emitting sounds having frequencies which differ only by a very small amount from each other. This results in a beat frequency, causing the variation from loud to soft, and back again.

□ □ □ □

11. Why does a vibrating guitar string sound louder when placed on the instrument than it would if allowed to vibrate in the air while off the instrument?

Answer

A vibrating string is not able to set very much air into motion when vibrated alone. Thus it will not be very loud. If it is placed on the instrument, however, the string's vibration sets the sounding board of the guitar into vibration. A vibrating piece of wood is able to move a lot of air, and the note is louder.

□ □ □ □

SOLUTIONS TO SELECTED END-OF-CHAPTER PROBLEMS

5. Two sinusoidal waves are described by

$$y_1 = (5.00 \text{ m})\sin[\pi(4.00x - 1200t)] \qquad y_2 = (5.00 \text{ m})\sin[\pi(4.00x - 1200t - 0.250)]$$

where x, y_1, and y_2 are in meters and t is in seconds. (a) What is the amplitude of the resultant wave? (b) What is the frequency of the resultant wave?

Solution

We can represent the waves symbolically as

$$y_1 = A_0 \sin(kx - \omega t)$$

and

$$y_2 = A_0 \sin(kx - \omega t - \phi)$$

with

$$A_0 = 5.00 \text{ m}, \qquad \omega = 1200\pi \text{ s}^{-1}, \qquad \text{and} \qquad \phi = 0.250\pi$$

According to the principle of superposition, the resultant wave function has the form

$$y = y_1 + y_2 = 2A_0 \cos\left(\frac{\phi}{2}\right)\sin\left(kx - \omega t - \frac{\phi}{2}\right)$$

(a) with amplitude

$$A = 2A_0 \cos\left(\frac{\phi}{2}\right) = 2(5.00)\cos\left(\frac{\pi}{8.00}\right) = 9.24 \text{ m} \qquad \lozenge$$

(b) and frequency

$$f = \frac{\omega}{2\pi} = \frac{1200\pi}{2\pi} = 600 \text{ Hz} \qquad \lozenge$$

11. Two speakers are driven in phase by a common oscillator at 800 Hz and face each other at a distance of 1.25 m. Locate the points along the line between the speakers where minima of the sound pressure amplitude would be expected. (Use $v = 343$ m/s.)

Solution The wavelength is

$$\lambda = \frac{v}{f} = \frac{343 \text{ m/s}}{800 \text{ Hz}} = 0.429 \text{ m}$$

$\leftarrow x \text{ m} \rightarrow \leftarrow (1.25 - x) \text{ m} \rightarrow$

The two waves moving in opposite directions along the line between the two speakers will add to produce a standing wave with this distance between nodes:

distance N to N = $\lambda/2 = 0.214$ m

Because the speakers vibrate in phase, air compressions from each will simultaneously reach the point halfway between the speakers, to produce an antinode of pressure here. A node of pressure will be located at this distance on either side of the midpoint:

distance N to A = $\lambda/4 = 0.107$ m

Therefore nodes of sound pressure will appear at these distances from either speaker:

$$\frac{1}{2}(1.25 \text{ m}) + 0.107 \text{ m} = 0.732 \text{ m}$$

$$\frac{1}{2}(1.25 \text{ m}) - 0.107 \text{ m} = 0.518 \text{ m}$$

The standing wave contains a chain of equally-spaced nodes at distances from either speaker of

0.732 m + 0.214 m = 0.947 m

0.947 m + 0.214 m = 1.16 m

1.16 m + 0.214 m = 1.38 m,   and so on,

and also at   0.518 m − 0.214 m = 0.303 m

0.303 m − 0.214 m = 0.0891 m

0.891 m − 0.214 m = −0.125 m,   etc.

In order, the distances from either speaker to the nodes of pressure between the speakers are at 0.0891 m, 0.303 m, 0.518 m, 0.732 m, 0.947 m, and 1.16 m

13. Two sinusoidal waves combining in a medium are described by the equations

$$y_1 = (3.0 \text{ cm})\sin \pi(x + 0.60t) \qquad\qquad y_2 = (3.0 \text{ cm})\sin \pi(x - 0.60t)$$

where x is in centimeters and t is in seconds. Determine the **maximum** displacement of the motion at (a) $x = 0.250$ cm, (b) $x = 0.500$ cm, and (c) $x = 1.50$ cm. (d) Find the three smallest values of x corresponding to antinodes.

Solution According to the waves in interference model, we add y_1 and y_2 using the trigonometry identity

$$\sin(\alpha + \beta) = \sin \alpha \cos \beta + \cos \alpha \sin \beta$$

We get

$$y = y_1 + y_2 = (6.0 \text{ cm})\sin(\pi x)\cos(0.60\pi t)$$

Since $\cos(0) = 1$, we can find the maximum value of y by setting $t = 0$:

$$y_{max}(x) = y_1 + y_2 = (6.0 \text{ cm})\sin(\pi x)$$

(a) At $x = 0.250$ cm, $\quad y_{max} = (6.0 \text{ cm})\sin(0.250\pi) = 4.24$ cm $\qquad\qquad \Diamond$

(b) At $x = 0.500$ cm, $\quad y_{max} = (6.0 \text{ cm})\sin(0.500\pi) = 6.00$ cm $\qquad\qquad \Diamond$

(c) At $x = 1.50$ cm, $\quad y_{max} = |(6.0 \text{ cm})\sin(1.50\pi)| = +6.00$ cm $\qquad\qquad \Diamond$

(d) The antinodes occur when $\qquad\qquad x = n\lambda/4 \quad (n = 1, 3, 5, \ldots)$

But $\quad k = 2\pi/\lambda = \pi$, so $\quad \lambda = 2.00$ cm, and $\quad x_1 = \lambda/4 = (2.00 \text{ cm})/4 = 0.500$ cm $\quad \Diamond$

$$x_2 = 3\lambda/4 = 3(2.00 \text{ cm})/4 = 1.50 \text{ cm} \quad \Diamond$$

$$x_3 = 5\lambda/4 = 5(2.00 \text{ cm})/4 = 2.50 \text{ cm} \quad \Diamond$$

17. Find the fundamental frequency and the next three frequencies that could cause standing wave patterns on a 30.0-m string that has a mass per length of 9.00×10^{-3} kg / m, and is stretched to a tension of 20.0 N.

Solution

The string described in the problem is very long, loose, and somewhat heavy, so it should have a very low fundamental frequency, maybe only a few vibrations per second.

The tension and linear density of the string can be used to find the wave speed, which can then be used along with the required wavelength to find the fundamental frequency.

The wave speed is

$$v = \sqrt{\frac{T}{\mu}} = \sqrt{\frac{20.0 \text{ N}}{9.00 \times 10^{-3} \text{ kg / m}}} = 47.1 \text{ m / s}$$

We use the wave under boundary conditions model. For a vibrating string of length L fixed at both ends, the wavelength of the fundamental is $\lambda = 2L = 60.0$ m; and the frequency is

$$f_1 = \frac{v}{\lambda} = \frac{v}{2L} = \frac{47.1 \text{ m / s}}{60.0 \text{ m}} = 0.786 \text{ Hz} \qquad \Diamond$$

The next three harmonics are

$$f_2 = 2f_1 = 1.57 \text{ Hz}, \ f_3 = 3f_1 = 2.36 \text{ Hz}, \ f_4 = 4f_1 = 3.14 \text{ Hz} \qquad \Diamond$$

The fundamental frequency is even lower than expected, less than 1 Hz. You could watch the string vibrating. It would weakly broadcast sound into the surrounding air, but all four of the lowest resonant frequencies are below the normal human hearing range (20 to 17000 Hz), so the sounds are not even audible.

21. A cello A-string vibrates in its first normal mode with a frequency of 220 vibrations/s. The vibrating segment is 70.0 cm long and has a mass of 1.20 g. (a) Find the tension in the string. (b) Determine the frequency of vibration when the string vibrates in three segments.

Solution

The length of the string $L = 0.700$ m, so the fundamental harmonic wavelength is

$$\lambda = 2L = 1.40 \text{ m}$$

and the velocity is

$$v = f\lambda = (220 \text{ s}^{-1})(1.40 \text{ m}) = 308 \text{ m/s}$$

(a) From the tension equation,

$$v = \sqrt{\frac{T}{\mu}} = \sqrt{\frac{T}{m/L}}$$

we calculate

$$T = \frac{v^2 m}{L} = \frac{(308 \text{ m/s})^2 (1.20 \times 10^{-3} \text{ kg})}{0.700 \text{ m}} = 163 \text{ N} \qquad \lozenge$$

(b) For the third harmonic state, the tension, linear density, and speed are the same. However, the string vibrates in three segments and the wavelength is one third as long as in the fundamental.

$$\lambda_3 = \lambda/3$$

From the equation $\lambda f = v$, we find that the frequency is three times as high:

$$f_3 = \frac{v}{\lambda_3} = 3\frac{v}{\lambda} = 3f = 660 \text{ Hz} \qquad \lozenge$$

25. Calculate the length of a pipe that has a fundamental frequency of 240 Hz if the pipe is (a) closed at one end and (b) open at both ends.

Solution

The relationship between the frequency and the wavelength of a sound wave is

$$v = f\lambda \qquad \text{or} \qquad \lambda = v/f$$

Next, we draw the pipes to help us visualize the relationship between λ and L.

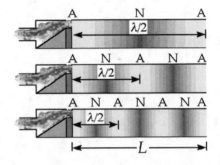

(a) For the fundamental mode in a closed pipe, $\qquad \lambda = 4L$

So $\qquad L = \dfrac{\lambda}{4} = \dfrac{v/f}{4}$ $\qquad$ and $\qquad L = \dfrac{(343 \text{ m / s})/(240 \text{ s}^{-1})}{4} = 0.357 \text{ m}$ ◊

(b) For the fundamental mode in an open pipe, $\qquad \lambda = 2L$

so $\qquad L = \dfrac{\lambda}{2} = \dfrac{v/f}{2}$ $\qquad$ and $\qquad L = \dfrac{(343 \text{ m / s})/(240 \text{ s}^{-1})}{2} = 0.715 \text{ m}$ ◊

31. A shower stall measures 86.0 cm × 86.0 cm × 210 cm. If you were singing in this shower, which frequencies would sound the richest (because of resonance)? Assume that the stall acts as a pipe closed at both ends, with nodes at opposite sides. Assume that the voices of the various singers range from 130 Hz to 2000 Hz. Let the speed of sound in the hot shower stall be 355 m/s.

Solution   For a closed box, the resonant frequencies will have nodes at both sides, so the permitted wavelengths will be defined by

$$L = \frac{n\lambda}{2} = \frac{nv}{2f}, \quad \text{with} \quad n = 1, 2, 3, \ldots$$

Rearranging, and substituting $L = 0.860$ m, the side-to-side resonant frequencies are

$$f_n = n\frac{v}{2L} = n\frac{355 \text{ m/s}}{2(0.860 \text{ m})} = n(206 \text{ Hz}), \quad \text{for each } n \text{ from 1 to 9} \qquad \lozenge$$

With $L' = 2.10$ m,   the top-to-bottom resonance frequencies are

$$f_n = n\frac{355 \text{ m/s}}{2(2.10 \text{ m})} = n(84.5 \text{ Hz}), \quad \text{for each } n \text{ from 2 to 23} \qquad \lozenge$$

----

33.   A glass tube is open at one end and closed at the other by a movable piston. The tube is filled with air warmer than that at room temperature, and a 384-Hz tuning fork is held at the open end. Resonance is heard when the piston is 22.8 cm from the open end and again when it is 68.3 cm from the open end. (a) What speed of sound is implied by these data? (b) How far from the open end will the piston be when the next resonance is heard?

$f = 384$ Hz

Warm air L

Solution   For an air column closed at one end, resonances will occur when the length of the column is equal to $\lambda/4$, $3\lambda/4$, $5\lambda/4$, and so on. Thus, the change in the length of the pipe from one resonance to the next is $\lambda/2$. In this case,

$$\lambda/2 = (0.683 \text{ m} - 0.228 \text{ m}) = 0.455 \text{ m} \qquad \text{and} \qquad \lambda = 0.910 \text{ m}$$

(a) $v = f\lambda = (384 \text{ Hz})(0.910 \text{ m}) = 349 \text{ m / s}$ $\qquad\qquad\qquad\qquad\qquad\qquad\qquad$ $\lozenge$

(b) $L = 0.683 \text{ m} + 0.455 \text{ m} = 1.14 \text{ m}$ $\qquad\qquad\qquad\qquad\qquad\qquad\qquad\qquad\qquad$ $\lozenge$

----

35. In certain ranges of a piano keyboard, more than one string is tuned to the same note to provide extra loudness. For example, the note at 110 Hz has two strings at this frequency. If one string slips from its normal tension of 600 N to 540 N, what beat frequency will be heard when the hammer strikes two strings simultaneously?

Solution   Combining the velocity equation $v = f\lambda$ and the tension equation $v = \sqrt{T/\mu}$

we find that

$$f = \sqrt{T/\mu\lambda^2}$$

and since μ and λ are constant,

$$\frac{f_1}{f_2} = \sqrt{\frac{T_1}{T_2}}$$

We solve

$$f_2 = (110 \text{ Hz})\sqrt{\frac{540 \text{ N}}{600 \text{ N}}} = 104.36 \text{ Hz}$$

and calculate the beat frequency

$$f_b = |f_1 - f_2| = 110 \text{ Hz} - 104.36 \text{ Hz} = 5.64 \text{ beats / s} \quad \Diamond$$

---

37. A student holds a tuning fork oscillating at 256 Hz. He walks toward a wall at a constant speed of 1.33 m/s. (a) What beat frequency does he observe between the tuning fork and its echo? (b) How fast must he walk away from the wall to observe a beat frequency of 5.00 Hz?

Solution   For an echo, $f_1 = f_2 \dfrac{v + v_s}{v - v_s}$ and the beat frequency $f_b = |f_1 - f_2|$

Solving for f_b gives $f_b = f_2 \dfrac{2v_s}{v - v_s}$ when approaching the wall.

(a) $f_b = (256 \text{ Hz})\dfrac{2(1.33 \text{ m/s})}{343 \text{ m/s} - 1.33 \text{ m/s}} = 1.99 \text{ Hz}$ $\Diamond$

(b)   When moving away from wall, v_s changes sign. Solving for v_s gives

$$v_s = f_b \frac{v}{2f_2 - f_b} = (5.00 \text{ Hz})\frac{343 \text{ m/s}}{2(256 \text{ Hz}) - 5.00 \text{ Hz}} = 3.38 \text{ m/s} \qquad \Diamond$$

---

51. A standing wave is set up in a string of variable length and tension by a vibrator of variable frequency. Both ends of the string are fixed. When the vibrator has a frequency f in a string of length L and tension T, n antinodes are set up in the string. (a) If the length of the string is doubled, by what factor should the frequency be changed so that the same number of antinodes is produced? (b) If the frequency and length are held constant, what tension will produce $n + 1$ antinodes? (c) If the frequency is tripled and the length of the string is halved, by what factor should the tension be changed so that twice as many antinodes are produced?

Solution Combining the equations $v = f\lambda$ and $v = \sqrt{T/\mu}$ and noting that $\lambda_n = 2L/n$

(a) we find that
$$f_n = \frac{n}{2L}\sqrt{\frac{T}{\mu}} \tag{1}$$

Keeping n, T, and μ constant, we can create two equations:

$$f_n L = \frac{n}{2}\sqrt{\frac{T}{\mu}} \qquad \text{and} \qquad f_n' L' = \frac{n}{2}\sqrt{\frac{T}{\mu}}$$

Dividing the equations, we find $\dfrac{f_n}{f_n'} = \dfrac{L'}{L}$

If $L' = 2L$, then $f_n' = \tfrac{1}{2}f_n$

Therefore, in order to double the length but keep the same number of antinodes, the frequency should be halved.   ◊

(b) From the same equation (1), we can hold L and f_n constant to get $\dfrac{n'}{n} = \sqrt{\dfrac{T}{T'}}$

From this relation, we see that the tension must be decreased to $T' = T\left(\dfrac{n}{n+1}\right)^2$

to produce $n + 1$ antinodes.   ◊

(c) This time, we rearrange equation (1) to produce $\dfrac{2f_n L}{n} = \sqrt{\dfrac{T}{\mu}}$ and $\dfrac{2f_n' L'}{n'} = \sqrt{\dfrac{T'}{\mu}}$

Dividing, we get $\dfrac{T'}{T} = \left(\dfrac{f_n'}{f_n}\cdot\dfrac{n}{n'}\cdot\dfrac{L'}{L}\right)^2 = \left(\dfrac{3f}{f_n}\right)^2\left(\dfrac{n}{2n}\right)^2\left(\dfrac{L/2}{L}\right)^2 = \dfrac{9}{16}$ ◊

53. If two adjacent natural frequencies of an organ pipe are determined to be 550 Hz and 650 Hz, calculate the fundamental frequency and length of this pipe. (Use $v = 340$ m/s.)

Solution   We use the wave under boundary conditions model.

Because harmonic frequencies are given by $f_o n$ for open pipes, and $f_o(2n - 1)$ for closed pipes, the difference between all adjacent harmonics is constant. Therefore, we can find each harmonic below 650 Hz by subtracting

$$\Delta f_{\text{Harmonic}} = (650 \text{ Hz} - 550 \text{ Hz}) = 100 \text{ Hz}$$

from the previous value.

The harmonics are: {650 Hz, 550 Hz, 450 Hz, 350 Hz, 250 Hz, 150 Hz, and 50 Hz}

(a)   The fundamental frequency, then, is 50 Hz.   ◊

(b)   The wavelength of the fundamental frequency can be calculated from the velocity:

$$\lambda = \frac{v}{f} = \frac{340 \text{ m/s}}{50.0 \text{ Hz}} = 6.80 \text{ m}$$

Because the step size Δf is twice the fundamental frequency, we know the pipe is closed, with an antinode at the open end, and a node at the closed end. The wavelength in this situation is four times the pipe length, so

$$L = \frac{\lambda}{4} = 1.70 \text{ m}$$ ◊

Chapter 15

Fluid Mechanics

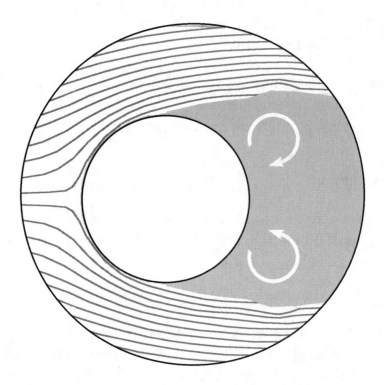

INTRODUCTION

In our treatment of the mechanics of fluids (liquids and gases), we shall see that Newton's laws and the work-kinetic energy theorem are sufficient to explain such effects as the buoyant force on a submerged object and the dynamic lift on an airplane wing. First, we consider a fluid at rest and derive an expression for the pressure exerted by the fluid as a function of its density and depth. We then treat fluids in motion, an area of study called **fluid dynamics**. A fluid in motion can be described by a model in which certain simplifying assumptions are made. We use this model to analyze some situations of practical importance. An underlying principle known as the **Bernoulli principle** enables us to determine relationships among the pressure, density, and speed at every point in a fluid. As we shall see, the Bernoulli principle is a result of conservation of energy applied to an ideal fluid.

NOTES FROM SELECTED CHAPTER SECTIONS

15.2   Variation of Pressure with Depth

In a **fluid at rest**, all points at the same depth are at the same pressure. **Pascal's law** states that a change in pressure applied to an **enclosed** fluid is transmitted undiminished to every point in the fluid and the walls of the containing vessel.

15.3   Pressure Measurements

The **absolute pressure** of a fluid is the sum of the **gauge pressure** and atmospheric pressure. The SI unit of pressure is the Pascal (Pa). Note that $1 \text{ Pa} \equiv 1 \text{ N/m}^2$.

15.4   Buoyant Forces and Archimedes's Principle

Any object partially or completely submerged in a fluid experiences a buoyant force equal in magnitude to the weight of the fluid displaced by the object and acting vertically upward through the point which was the center of gravity of the displaced fluid.

15.5   Fluid Dynamics

When fluid is in motion, its flow can be characterized as being one of two main types. The flow is said to be steady if each particle of the fluid follows a smooth path, and the paths of each particle do not cross each other. Above a certain critical speed, fluid flow becomes nonsteady or turbulent, characterized by small whirlpool-like regions.

The term **viscosity** is commonly used in fluid flow to characterize the degree of internal friction in the fluid. This internal friction is associated with the resistance to two adjacent layers of the fluid to move relative to each other. Because of viscosity, part of the kinetic energy of a fluid is converted to internal energy.

In our model of an ideal fluid, we make the following four assumptions:

- **Nonviscous fluid.** In a nonviscous fluid, internal friction is neglected. An object moving through a nonviscous fluid would experience no retarding viscous force.

- **Steady flow.** In a steady flow, we assume that the velocity of the fluid at each point remains constant in time.

- **Incompressible fluid.** The density of the fluid is assumed to remain constant in time.

- **Irrotational flow.** Fluid flow is irrotational if there is no angular momentum of the fluid about any point. If a small wheel placed anywhere in the fluid does not rotate about its center of mass, the flow would be considered irrotational. (If the wheel were to rotate, as it would if turbulence were present, the flow would be rotational.)

15.6 Streamlines and the Continuity Equation in Fluids

The path taken by a fluid particle under **steady flow** is called a **streamline**. The velocity of the fluid particle is always tangent to the streamline at that point and no two streamlines can cross each other. A set of streamlines forms a **tube of flow**. In steady flow, particles of the fluid cannot flow into or out of a tube of flow.

EQUATIONS AND CONCEPTS

The **density** of a homogeneous substance is defined as its ratio of mass to volume. The value of density is characteristic of a particular type of material and independent of the total quantity of material in the sample.

$$\rho \equiv \frac{m}{V}$$

The SI units of density are kg per cubic meter.

$$1\,g/cm^3 = 1000\,kg/m^3$$

The (average) pressure of a fluid is defined as the normal force per unit area acting on a surface immersed in the fluid.

$$P \equiv \frac{F}{A}$$

(15.1)

The SI units of pressure are newtons per square meter, or Pascal (Pa).

$$1\,Pa \equiv 1\,N/m^2$$

(15.2)

Atmospheric pressure is often expressed in other units: atmospheres, mm of mercury (Torr), or pounds per square inch.

$$P_0 = 1\,atm \approx 1.013 \times 10^5\,Pa$$

$$1\,Torr = 133.3\,Pa$$

$$1\,lb/in^2 = 6.89 \times 10^3\,Pa$$

The absolute pressure, P, at a depth, h, below the surface of a liquid which is open to the atmosphere is greater than atmospheric pressure, P_0, by an amount which depends on the depth below the surface.

$$P = P_0 + \rho g h$$

(15.4)

The quantity $P_g = \rho gh$ is called the gauge pressure and P is the absolute pressure. Therefore,

$$P_{absolute} = P_{atmosphere} + P_g$$

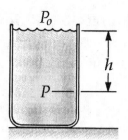

Pascal's law states that pressure applied to an enclosed fluid (liquid or gas) is transmitted undiminished to every point within the fluid and to the walls of the vessel which contain the fluid.

Archimedes's principle states that when an object is partially or fully submerged in a fluid, the fluid exerts an upward buoyant force on the object. The magnitude of the buoyant force depends on the density of the fluid and the volume of displaced fluid, V. In particular, note that B equals the weight of the displaced fluid.

$$B = \rho_f g V \qquad (15.5)$$

For a floating object, the fraction of the object under the fluid surface equals the ratio of the object density to the fluid density.

$$\frac{\rho_0}{\rho_f} = \frac{V}{V_0} \qquad (15.6)$$

The equation of continuity for fluids is a statement of conservation of mass: **the mass that flows into a tube equals the mass that flows out.**

For an **incompressible fluid** (ρ = constant), the equation of continuity can be written as Equation 15.7. The product Av is called the volume flow rate. Therefore, the flow rate at any point along a pipe carrying an incompressible fluid is constant.

$$\rho_1 A_1 v_1 = \rho_2 A_2 v_2$$

$$A_1 v_1 = A_2 v_2 \qquad (15.7)$$

Bernoulli's equation is a fundamental law in fluid mechanics. The equation is a statement of the law of conservation of mechanical energy as applied to a fluid. Bernoulli's equation states that the sum of pressure, kinetic energy per unit volume, and gravitational potential energy per unit volume remains constant throughout an ideal fluid.

$$P + \tfrac{1}{2}\rho v^2 + \rho g y = \text{constant} \qquad (15.13)$$

REVIEW CHECKLIST

▷ Understand the concept of pressure at a point in a fluid, and the variation of pressure with depth. Understand the relationships among absolute, gauge, and atmospheric pressure values; and know the several different units commonly used to express pressure.

▷ Understand the origin of buoyant forces; and state and explain Archimedes's principle.

▷ State the simplifying assumptions of an ideal fluid moving with streamline flow.

▷ State and understand the physical significance of the **equation of continuity** (constant flow rate) and **Bernoulli's equation** for fluid flow (relating **flow velocity, pressure, and pipe elevation**).

ANSWERS TO SELECTED CONCEPTUAL QUESTIONS

4. A barge is carrying a load of gravel along a river. It approaches a low bridge, and the captain realizes that the top of the pile of gravel is not going to make it under the bridge. The captain orders the crew to quickly shovel gravel from the pile into the water. Is this a good decision?

Answer It turns out that this is not a good decision. Imagine a given shovel full of gravel. According to Archimedes's principle, the portion of the total buoyant force which can be associated with keeping this shovel-full of gravel afloat on the barge is due to a portion of water whose weight is equal to the weight of the gravel. Since water is less dense than gravel, the volume of water associated with keeping a shovel-full of gravel afloat is larger than the volume of the gravel. When the gravel is thrown overboard, the barge rises, in association with a volume of water larger than the volume of the gravel removed from the pile. Thus, as the gravel is shoveled overboard, the barge rises by an amount which is larger than the amount by which the height of the pile of gravel decreases. The better approach would be to **add** gravel to the barge from a riverside source, causing it to sink deeper into the water at a greater rate than the height of the pile of gravel increases.

6. Two drinking glasses having equal weights but different shapes and different cross-sectional areas are filled to the same level with water. According to the expression $P = P_0 + \rho gh$, the pressure is the same at the bottom of both glasses. In view of this, why does one weigh more than the other?

Answer For the cylindrical container shown, the weight of the water is equal to the gauge pressure at the bottom multiplied by the area of the bottom. For the narrow-necked bottle, on the other hand, the weight of the fluid is much less than PA, the bottom pressure times the area. The water does exert the large PA force downward on the bottom of the bottle, but the water also exerts an upward force nearly as large on the ring-shaped horizontal area surrounding the neck. The net vector force that the water exerts on the glass is equal to the small weight of the water.

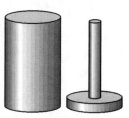

8.   A fish rests on the bottom of a bucket of water while the bucket is being weighed. When the fish begins to swim around, does the weight change?

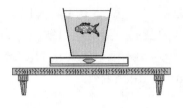

Answer   In either case, the scale is supporting the container, the water, and the fish. Therefore the weight will remains the same. The reading on the scale, however, can change if the net center of mass accelerates in the vertical direction, as when the fish jumps out of the water. In that case, the scale, which registers force, will also show an additional force caused by Newton's law, $F = ma$.

□   □   □   □

12.   The water supply for a city is often provided from reservoirs built on high ground. Water flows from the reservoir, through pipes, and into your home when you turn the tap on your faucet. Why is the water flow more rapid out of a faucet on the first floor of a building than in an apartment on a higher floor?

Answer   The water supplied to the building flows through a pipe connected to the water tower. Near the earth, the water pressure is greater because the pressure increases with increasing depth beneath the surface of the water. The penthouse apartment is not as far below the water surface, hence the water flow will not be as rapid as in a lower floor.

□   □   □   □

23.   An unopened can of diet cola floats when placed in a tank of water, whereas a can of regular cola of the same brand sinks in the tank. What do you suppose could explain this behavior?

Answer   Regular cola is sugar syrup. Its density is higher than the density of diet cola, which is nearly pure water. The low-density air inside the can has a bigger effect than the thin aluminum shell, so the can of diet soda floats.

□   □   □   □

SOLUTIONS TO SELECTED END-OF-CHAPTER PROBLEMS

1. Calculate the mass of a solid iron sphere that has a diameter of 3.00 cm.

Solution   The definition of density $\rho = m/V$ is often written as $m = \rho V$.

Here $V = \frac{4}{3}\pi r^3$ so $m = \rho V = \rho\left(\frac{4}{3}\pi(d/2)^3\right)$

Thus, $m = \left(7.86 \times 10^3 \text{ kg}/\text{m}^3\right)\left(\frac{4}{3}\right)\left((1.50 \text{ cm})^3\right)\left(10^{-6} \text{ m}^3/\text{cm}^3\right) = 111 \text{ g}$ ◊

3. A 50.0-kg woman balances on one heel of a pair of high-heeled shoes. If the heel is circular and has a radius of 0.500 cm, what pressure does she exert on the floor?

Solution   The area of the circular base of the heel is

$$\pi r^2 = \pi(0.500 \text{ cm})^2\left(\frac{1 \text{ m}^2}{10\ 000 \text{ cm}^2}\right) = 7.85 \times 10^{-5} \text{ m}^2$$

The force she exerts is her weight, $mg = (50.0 \text{ kg})(9.80 \text{ m}/\text{s}^2) = 490 \text{ N}$

Then $P = \dfrac{F}{A} = \dfrac{490 \text{ N}}{7.85 \times 10^{-5} \text{ m}^2} = 6.24 \text{ MPa}$ ◊

7. The spring of the pressure gauge shown in Figure 15.2 has a force constant of 1000 N/m, and the piston has a diameter of 2.00 cm. As the gauge is lowered into water, what change in depth causes the piston to move in by 0.500 cm?

Solution $F_{\text{spring}} = F_{\text{fluid}}$ or $kx = \rho g h A$

$h = \dfrac{kx}{\rho g A} = \dfrac{\left(1000 \text{ N}/\text{m}^2\right)(0.00500 \text{ m})}{\left(1000 \text{ kg}/\text{m}^3\right)\left(9.80 \text{ m}/\text{s}^2\right)(0.0100 \text{ m})^2 \pi} = 1.62 \text{ m}$

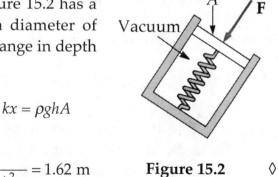

Figure 15.2   ◊

9. What must be the contact area between a suction cup (completely exhausted) and a ceiling if the cup is to support the weight of an 80.0-kg student?

Solution "Suction" is not a new kind of force. Familiar forces hold the cup in equilibrium. The ceiling pushes down on the cup with some normal force. The student pulls down with a force of magnitude

$$F_g = mg = (80.0 \text{ kg})(9.80 \text{ m/s}^2) = 784 \text{ N}$$

The vacuum between cup and ceiling exerts no force on either. The air below the cup pushes up on it with a force $(P_{atm})(A)$. If the cup barely supports the student, the normal force of the ceiling is nearly zero, and the particle in equilibrium model gives

$$\Sigma F_y = 0 + P_{atm}A - mg = 0$$

$$A = \frac{mg}{P_{atm}} = \frac{784 \text{ N}}{1.013 \times 10^5 \text{ N/m}^2} = 7.74 \times 10^{-3} \text{ m}^2$$

$\Diamond$

---

13. Blaise Pascal duplicated Torricelli's barometer using a red Bordeaux wine of density 984 kg/m^3 as the working liquid (Fig. P15.13). What was the height h of the wine column for normal atmospheric pressure? Would you expect the vacuum above the column to be as good as for mercury?

Solution In Bernoulli's equation,

$$P_1 + \tfrac{1}{2}\rho v_1^2 + \rho g y_1 = P_2 + \tfrac{1}{2}\rho v_2^2 + \rho g y_2$$

Take point 1 at the wine surface in the pan, where $P_1 = P_{atm}$, and point 2 at the wine surface up in the tube. Here we approximate $P_2 = 0$, although some alcohol and water will evaporate. The vacuum is not as good as with mercury. Unless you are careful, a lot of dissolved oxygen or carbon dioxide may come bubbling out.

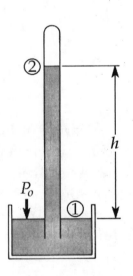

Figure P15.13
(modified)

Now, since the speed of the fluid at both points is zero, we can rearrange the terms to arrive essentially at Equation 15.4:

$$P_1 = P_2 + \rho g(y_2 - y_1)$$

$$1 \text{ atm} = 0 + (984 \text{ kg/m}^3)(9.80 \text{ m/s}^2)(y_2 - y_1)$$

$$y_2 - y_1 = \frac{1.013 \times 10^5 \text{ N/m}^2}{9643 \text{ N/m}^3} = 10.5 \text{ m} \qquad \Diamond$$

A water barometer in a stairway of a three-story building is a nice display. Red wine makes the fluid level easier to see.

---

17. A Ping-Pong ball has a diameter of 3.80 cm and average density of 0.0840 g/cm³. What force is required to hold it completely submerged under water?

Solution

At equilibrium, where B is the buoyant force,

$$\Sigma F = 0 \qquad \text{or} \qquad F_{app} + mg - B = 0$$

The applied force, $\quad F_{app} = B - mg \quad$ when $B = V \rho_w g$

Since $m = V \rho_{ball}$, $\quad F_{app} = V g(\rho_w - \rho_{ball}) = \frac{4}{3}\pi r^3 g(\rho_w - \rho_{ball})$

$$F_{app} = \frac{4}{3}\pi(1.90 \times 10^{-2} \text{ m})^3 (9.80 \text{ m/s}^2)(1000 \text{ kg/m}^3 - 84.0 \text{ kg/m}^3)$$

$$F_{app} = 0.258 \text{ N} \qquad \Diamond$$

---

21. A cube of wood having an edge dimension of 20.0 cm and a density of 650 kg/m³ floats on water. (a) What is the distance from the horizontal top surface of the cube to the water level? (b) How much lead weight has to be placed on top of the cube so that its top is just level with the water?

Solution   Set h equal to the distance from the top of the cube to the water level.

(a) According to Archimedes's principle,

$$B = \rho_w V g = \left(1.00 \text{ g} / \text{cm}^3\right)\left[(20.0 \text{ cm})^2(20.0 \text{ cm} - h \text{ cm})\right]g$$

But $B = \text{Weight of block} = mg = \rho_{wood} V_{wood} g = \left(0.650 \text{ g} / \text{cm}^3\right)(20.0 \text{ cm})^3 g$

Setting these two equations equal,

$$(0.650 \text{ g}/\text{cm}^3)(20.0 \text{ cm})^3 g = (1.00 \text{ g}/\text{cm}^3)(20.0 \text{ cm})^2(20.0 \text{ cm} - h \text{ cm}) \, g$$

$$20.0 - h = 20.0(0.650) \qquad\qquad h = 20.0(1.00 \text{ cm} - 0.650 \text{ cm}) = 7.00 \text{ cm} \qquad ◊$$

(b) $B = mg + Mg$ where $M = \text{mass of lead}$

$$(1.00 \text{ g}/\text{cm}^3)(20.0 \text{ cm})^3 g = (0.650 \text{ g}/\text{cm}^3)(20.0 \text{ cm})^3 g + Mg$$

$$M = (20.0 \text{ cm})^3\left(1.00 \text{ g} / \text{cm}^3 - 0.650 \text{ g} / \text{cm}^3\right) = (20.0 \text{ cm})^3\left(0.350 \text{ g} / \text{cm}^3\right) = 2.80 \text{ kg} \qquad ◊$$

---

23. How many cubic meters of helium are required to lift a balloon with a 400-kg payload to a height of 8000 m? (Take $\rho_{He} = 0.180 \text{ kg}/\text{m}^3$.) Assume that the balloon maintains a constant volume and that the density of air decreases with altitude z according to the expression $\rho_{air} = \rho_0 e^{-z/8000}$, where z is in meters, and $\rho_0 = 1.25 \text{ kg}/\text{m}^3$ is the density of air at sea level.

Solution   At $z = 8000$ m, the density of air is

$$\rho_{air} = \rho_0 e^{-z/8000} = \left(1.25 \text{ kg}/\text{m}^3\right)e^{-1} = \left(1.25 \text{ kg}/\text{m}^3\right)(0.368) = 0.460 \text{ kg}/\text{m}^3$$

Think of the balloon reaching equilibrium at this height.

The weight of its payload is $\qquad\qquad Mg = (400\ \text{kg})(9.80\ \text{m/s}^2) = 3920\ \text{N}$

The weight of the helium in it is $\qquad mg = \rho_{He}Vg$

$\sum F_y = 0$ becomes $\qquad\qquad\qquad +\rho_{air}Vg - Mg - \rho_{He}Vg = 0$

Solving, $\qquad\qquad\qquad\qquad\qquad (\rho_{air} - \rho_{He})V = M$

and $\qquad\qquad\qquad\qquad\qquad V = \dfrac{M}{\rho_{air} - \rho_{He}} = \dfrac{400\ \text{kg}}{(0.460 - 0.18)\ \text{kg/m}^3} = 1.43 \times 10^3\ \text{m}^3 \qquad \Diamond$

---

25. A plastic sphere floats in water with 50.0% of its volume submerged. This same sphere floats in glycerin with 40.0% of its volume submerged. Determine the densities of the glycerin and the sphere.

Solution

The forces on the ball are its weight $\qquad F_g = mg = \rho_{plastic}V_{ball}g$

and the buoyant force of the liquid $\qquad B = \rho_{fluid}V_{immersed}g$

When floating in water, $\sum F_y = 0$: $\qquad -\left(\rho_{plastic}\right)\left(V_{ball}\right)g + \left(\rho_{water}\right)\left(0.500V_{ball}\right)g = 0$

$$\rho_{plastic} = 0.500\rho_{water} = 500\ \text{kg/m}^3 \qquad \Diamond$$

When floating in glycerin, $\sum F_y = 0$: $\qquad -\left(\rho_{plastic}\right)\left(V_{ball}\right)g + \left(\rho_{glycerine}\right)\left(0.400V_{ball}\right)g = 0$

$$\rho_{plastic} = 0.400\rho_{glycerine}$$

$$\rho_{glycerine} = \dfrac{500\ \text{kg/m}^3}{0.400} = 1250\ \text{kg/m}^3 \qquad \Diamond$$

This glycerin would sink in water.

---

31. A large storage tank with an open top is filled to a height h_0. The tank is punctured at a height h above the bottom of the tank (Fig. P15.31). Find an expression for how far from the tank the exiting stream lands.

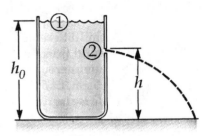

Figure P15.31
(modified)

Solution

Take point 1 at the top liquid surface. Since the tank is large, the fluid level falls only slowly and $v_1 \cong 0$. Take point 2 at the surface of the water stream leaving the hole. At both points the pressure is one atmosphere because the water can push no more or less strongly than the air pushes on it, as described by Newton's third law.

$$P_1 + \tfrac{1}{2}\rho v_1^2 + \rho g y_1 = P_2 + \tfrac{1}{2}\rho v_2^2 + \rho g y_2$$

$$P_a + 0 + \rho g h_0 = P_a + \tfrac{1}{2}\rho v_2^2 + \rho g h$$

$$v_2 = \sqrt{2g(h_0 - h)}$$

Now each drop of water moves as a projectile. Its velocity v_2, which we now call its original velocity, has zero vertical component. Its time of fall is given by the particle under constant acceleration model, from $y = v_{y0}t + \tfrac{1}{2}a_y t^2$:

$$-h = 0 - \tfrac{1}{2}gt^2 \qquad \text{so} \qquad t = \sqrt{\frac{2h}{g}}$$

and from the particle under constant velocity model its horizontal displacement is

$$x = v_{x0}t + \tfrac{1}{2}a_x t^2 = \sqrt{2g(h_0 - h)}\sqrt{2h/g} + 0 = \sqrt{4h(h_0 - h)} \qquad \lozenge$$

45. The true weight of an object is measured in a vacuum, where buoyant forces are absent. A body of volume V is weighed in air on a balance with the use of weights of density ρ . If the density of air is ρ_{air} and the balance reads F'_g, show that the true weight F_g is

$$F_g = F'_g + \left(V - \frac{F'_g}{\rho g}\right)\rho_{air} g$$

Solution We use the rigid body in equilibrium model. The "balanced" condition is one in which the net torque on the balance is zero. Since the balance has lever arms of equal length, the total force on each pan is equal. Applying $\Sigma\tau = 0$ around the pivot gives us this in equation form:

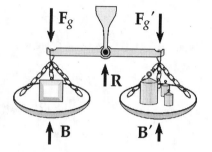

$$F_g - B = F'_g - B'$$

where B and B' are the buoyant forces on the body and weights respectively. The buoyant force experienced by an object of volume V in air is $B = V\rho_{air} g$. So for the test mass and for the weights, respectively,

$$B = V\rho_{air} g \qquad \text{and} \qquad B' = V'\rho_{air} g$$

Since the volume of the weights is not given explicitly, we must use the density equation to eliminate it:

$$V' = \frac{m'}{\rho} = \frac{m'g}{\rho g} = \frac{F'_g}{\rho g}$$

With this substitution, the buoyant force on the weights is $B' = \left(F'_g / \rho g\right)\rho_{air} g$

Therefore, $\quad F_g = F'_g + \left(V - \dfrac{F'_g}{\rho g}\right)\rho_{air} g$ $\qquad\qquad\qquad\qquad\qquad\qquad \Diamond$

Related Comment: We can now answer the popular riddle: Which weighs more, a pound of feathers or a pound of bricks? Like in the problem above, the feathers have a greater buoyant force than the bricks, so if they "weigh" the same on a scale as a pound of bricks, then the feathers must have more mass and therefore a greater "true weight."

47. With reference to Figure 15.6, show that the total torque exerted by the water behind the dam about an axis through O is $\frac{1}{6}\rho g w H^3$. Show that the effective line of action of the total force exerted by the water is at a distance $\frac{1}{3}H$ above O.

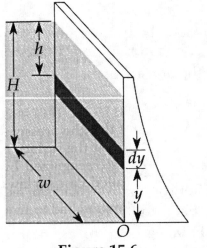

Figure 15.6

Solution   The torque is calculated from the equation

$$\tau = \int d\tau = \int r\,dF$$

From Figure 15.6, we have

$$\tau = \int (y)dF = \int_0^H y\big[\rho g(H-y)w\big]dy = \tfrac{1}{6}\rho g w H^3 \qquad \lozenge$$

$$F = \int dF = \int_0^H \big(\rho g(H-y)w\big)dy = \tfrac{1}{2}\rho g w H^2$$

If this were applied at a height y_{eff} such that the torque remains unchanged,

$$\tfrac{1}{6}\rho g w H^3 = y_{\text{eff}}\Big[\tfrac{1}{2}\rho g w H^2\Big] \qquad \text{and} \qquad y_{\text{eff}} = \tfrac{1}{3}H \qquad \lozenge$$